# SCIENCE AND THE
# SECOND RENAISSANCE
# OF EUROPE

Published for the Commission of the European Communities,
European Committee for Research and Development

# SCIENCE AND THE SECOND RENAISSANCE OF EUROPE

by

## A. DANZIN

Chairman of the European Committee for Research and Development (CERD)

Published for the

COMMISSION OF THE EUROPEAN COMMUNITIES

by

**PERGAMON PRESS**

OXFORD · NEW YORK · TORONTO · SYDNEY · PARIS · FRANKFURT

| | |
|---|---|
| U.K. | Pergamon Press Ltd., Headington Hill Hall, Oxford OX3 0BW, England |
| U.S.A. | Pergamon Press Inc., Maxwell House, Fairview Park, Elmsford, New York 10523, U.S.A. |
| CANADA | Pergamon of Canada, Suite 104, 150 Consumers Road, Willowdale, Ontario M2 J1P9, Canada |
| AUSTRALIA | Pergamon Press (Aust.) Pty. Ltd., P.O. Box 544, Potts Point, N.S.W. 2011, Australia |
| FRANCE | Pergamon Press SARL., 24 rue des Ecoles, 75240 Paris, Cedex 05, France |
| FEDERAL REPUBLIC OF GERMANY | Pergamon Press GmbH, 6242 Kronberg-Taunus, Pferdstrasse 1, Federal Republic of Germany |

Copyright © 1979 ECSC, EEC, EAEC, Luxembourg

*All Rights Reserved. No part of this publication may be reproduced, stored in a retrieval system or transmitted in any form or by any means: electronic, electrostatic, magnetic tape, mechanical, photocopying, recording or otherwise, without permission in writing from the copyright holders.*

First edition 1979

**British Library Cataloguing in Publication Data**

Danzin, A
Science and the second renaissance of Europe.
1. Technological innovations - European
Economic Community Countries
I. Title
609'.4          T173.8          78-40766
ISBN 0-08-022442-3

EUR 5982 ef

*Printed in Great Britain by Aberdeen University Press, Aberdeen*

# CONTENTS

PART II. Proposals for Mild Treatment:
A Community Scientific and Technical Policy

# FOREWORD

## by Prof. I. Prigogine

It is a particular pleasure for me to be writing this brief foreword to the
important essay by André Danzin. Not surprisingly, I was moved both as a
scientist and as a European by the problems raised. The roles and the very
existence of science and of Europe will henceforward be joined together, each
acting as a powerful influence on the other.

The reader will appreciate, as I did, André Danzin's clarity of perception as he
paints his picture, setting Europe in perspective. To read about this landscape
with him is to grasp the principal points of strength for evaluation of the
semistrategic course which must be taken.

The picture is undoubtedly realistic. Although the author stresses the particular
importance of science within the cultural, economic, and social fabric of Europe,
he does so to bring to the fore serious problems which lie on the periphery of what
must be referred to as a certain urgency. These problems are to some extent
paradoxical. On the one hand, I am firmly convinced that we are capable of
achieving the best. I see no noticeable difference here between young research
workers encountered in Europe and in the United States; it can only be said that
they have the same quality of training and enthusiasm, and I feel the same pleasure
in discussing their projects with them. However, whilst this may portend well for
the future, it must be admitted that the present shows signs of being a double
failure.

Applied European research is at present drastically dependent. We are already
paying the price, and shall pay more and more dearly in the future, for not granting
the investments required to set up industrial infrastructures geared to the growth
industries. Not only are we failing to reap the benefits of technological
achievements which reward, accompany, and stimulate investment, we are furthermore
obliged to subsidize the dynamism of those countries from whom we purchase and hire
products and services typical of these sectors, whether satellite vehicles or data
processing.

With regard to pure research, whatever the criterion adopted (mention in reviews, scientific esteem, etc.), European research looks as though it will find itself reduced to being an adjunct to a movement in which it is no longer a driving force.

I should not like these thoughts to be misconstrued.  I have the greatest admiration for the innovating dynamism of large countries like the United States. I am merely concerned for the future of research which must surely affect the development of Europe.

As a citizen of a small country, I understand very clearly the handicaps described by Danzin in his analysis.  Political partitioning of states and the burden of the past have so far prevented many of our scientific institutions from embarking upon specialized studies and achieving the breadth necessary to carry out meaningful projects.  At the same time the lack of mobility amongst research workers is depriving them of the specialism essential for any innovating research.

The solution to the problem (magnitude, diversification, mobility) implies reorganization on a European scale in view of the low level of centralization of the EEC's financial resources at the present time.

The value of Danzin's analyses lies in the fact that they lead on to solutions which take account of the resources and data inherent in the problems.  This is tantamount to saying that action can be taken right away by finding a few sensitive fields where there is a European social demand and where it is possible to engage small teams likely to mobilize advanced scientific resources in which Europe can cut a dash.  Such action might have — to use André Danzin's felicitous phrase — a catalytic effect on scientific advance and social progress in Europe.

Since I have spent a part of my scientific career studying fluctuations in science, I am sufficiently familiar with the extraordinary effect which a minor fluctuation, occurring where there is a certain instability, can have on the organization of the whole field of science.

"A second renaissance or acceptance of decline?" asks Danzin with striking insight. By improving our understanding of these problems he will contribute to the decision-making process which over the years to come will shape the future of Europe.

# FOREWORD

## By Dr. G. Brunner

Since in the next few decades, Europe will be confronted by many major complicated
problems, we must increase our efforts to find new solutions now.  The development
of a new world economic growth in the Community and the structural change which our
high-technology civilization is undergoing, all involve a vast number of problems.

Energy and raw material requirements, environmental protection, and the population
explosion in many developing countries are all critical factors which we cannot
ignore.  Science and technology will often have a leading role to play in meeting
these undeniable and unavoidable challenges.  The European public is aware of this
fact, and in a recent study carried out by the Commission of the European Communities it clearly expresses the hopes it pins on, and the confidence it has in, science
as the instrument which will make tomorrow's world a better place in which to live.
However, confidence in scientific progress as the vehicle for economic growth
and consequently for the welfare of the Community's citizens is no longer undisputed;
there are even those in the Member States who recommend a slowing down of scientific
and technological activity, fearing that its disadvantages will outweigh its
advantages.

Given this situation, the European Research and Development Committee discussions
have revealed the need for in-depth consideration of both the concept of the
European Communities and the function of science and technology in Europe.  The
Advisory Committee, comprising twenty-one members of known independence, assists
and advises the Commission in defining common policy on research and development.
We are greatly indebted to its Chairman, Mr. A. Danzin, for having provided a number of ideas for consideration in response to this need and for helping us to make
a clear analysis of our situation.

*Science and the Second Renaissance of Europe* is an important work for several
reasons.

First, it provides a survey which is an analysis of the present situation in Europe
(as regards demography, economics, technology, etc.) and demonstrates the vulnerability of a Europe where space is limited and the balance of the environment easily
upset, which lacks the primary resources essential for its development and which
will soon be facing serious demographic problems.

In spite of the handicaps from which it suffers, the European Community must continue its growth if it is to offer a response to the requirements and satisfy the

aspirations of the peoples of Europe.  This need and these aspirations are the subject of the second part of the book, which is really an attempt to define a new type of growth which takes account of both the "European situation" and the *idées-forces* of European culture.  The second section is therefore a plea for a new type of European development based on the strength of European culture and the abilities of the Europeans.  In this framework, emphasis would be placed on intellectual activities, on activities in the field of information (dissemination, processing), and on co-operation with the developing countries — all activities sparing in consumption of raw materials and energy and which have immense development potential.

If such a development model is to be contemplated, the facilities for creative R&D activities will have to be provided on a large scale.  Individual countries' programmes are inadequate, being too limited in terms of resources and experience.  In a third section, therefore, the author outlines a plan for science and technology in the European Community.  It is worth pointing out in this connection that there is a definite dovetailing of the aims of this project with those of the common scientific and technical policy which is already being pursued by the European Community.  The views propounded are of interest both as regards methods and procedures (budgetary control, "executive body", catalytic action) and as regards the proposed areas of activity.

They can be seen as an encouragement to proceed along the path we have already mapped out and a source of stimulating ideas for the constant improvement of Community strategy and methods of operation in the field of R&D.

The proposed plan for science and technology in the European Community sometimes appears too limited in view of the extent and complexity of the problems to be solved in order to meet the requirements and aspirations of the citizens of the Community.  But whether or not one accepts the proposed "scientific and technical programme" in its entirety, it represents a possible route to a new, a second Renaissance of Europe marked by progress along lines which would be in keeping with our Continent's specific genius.

I am therefore pleased that one of Europe's leading scientists, who also has considerable experience of industry, has provided us with this overall survey of our situation, our capacities, and our requirements, which will help us to examine the problem and lay the foundations for what I hope will be the broadest possible discussion of the question among the peoples of Europe.

# PREFACE

This book is the offspring of anguish and hope.

Europe may well slip towards underdevelopment.  This study shows that the risk is a
real one and that very little seems to be being done to counteract it.  Europe,
eroded by scepticism, lacking ambition, wanting in projects relating to man and
society, is not even aware that its divisions make it ill-suited to the dimensions of
the modern world.  This vision fosters anguish.

Hope comes from the European heritage and the forces of change that the future
requires.  Moral and cultural values are still alive and form the background for
challenge; even if they are discussed, their questioning may be viewed as the sign
of an effort to adapt, a move towards the renewal of progress, towards a new
Renaissance.  The world has never had such a great need of a social laboratory to
overcome the difficulties caused by the very excesses of human success.  Europe
is under pressure and appears predestined to become the melting-pot from which will
come a resurgence.

This meditation on Europe leads one to ask why, when so many other economic and
political factors could be regarded as more important and relevant, scientific and
technical innovation should be singled out for special consideration?  The answer
lies in a deeper line of thought; it arises from a new concept of evolution that is
acceptable today.

Even when they differ in their metaphysical views, scientists agree that evolution
in the living world is always in the same direction: the drive towards complexity,
the constant increase in mental processes.[1]  Throughout geological time the
apparent form of this progress has been, through the random action of genetic
mutations, a long series of trials and new solutions which only the best-adapted
forms have survived.  Although there has been no systematic destruction of ancient
species, in particular as a result of the conservation of lateral forms, the
success of each new species has generally caused that from which it was directly
derived to disappear.  In evolution it is fair to say, from several points of view,
that "life flourishes on death".

Since Man appeared, the transformation of the living world has been overtaken by the
speed of evolution of the artificial universe, generated by the human hand and brain;
evolution has moved to the field of culture and social life.  Genetically, man is
stabilized.  However, evolution continues with the accumulation of tools forming the

---

[1] Cf. Jacques Ruffié, *De la Biologie à la Culture*, Flammarion, 1977.

"machine kingdom",[1] which is on the way to supplanting the plant and animal
kingdoms to which nature was adapted.  This coexistence with our increasingly
powerful tools is bringing about a profound evolution in our society.  The drive
towards complexity and psychism continues, but it has become the work and
responsibility of the human race, and suddenly the process has speeded up.

And yet, confusing as it may seem, this evolution retains all the outward signs of
a gamble between chance and necessity.  The field in which chance comes into play
is that of scientific research; selection by necessity is carried out by economic
competition, by the development of innovation.[2]  The acknowledged role of chance
reopens from the philosophic viewpoint, the full range of problems arising from
relations between responsibility and freedom and all our ideas on planning
structures.

Although it is justifiable to link the changes in the present-day world with evolution,
the association must nevertheless be examined with the closest attention.  History
teaches us that the backward are liable to be eliminated at any moment, and to the
selection process man has added a ruthless cruelty.  Today the conflict between the
societies frozen at the pre-industrial stage and the mutated post-industrial
societies shows a dual imbalance.  The overwhelming population pressure amongst the
poor peoples is accompanied by the crushing weight of the arms and economic power
of the rich, rich essentially by reason of their technological superiority.  There
is every indication that over the next quarter of a century the main mode of
selection will continue to be the power of technological innovation.

Man refuses, however, to submit to a purely biological law of evolution; he is
aware of his responsibility.  A great step forward in recent years has been the
realization that there are certain limits to man's power over his destiny, despite
the powerfulness of the new tools; science knows that it will not be able to
explain everything nor to govern everything.  This return to a healthy modesty
does not rule out responsibility; man can choose the fields in which to apply his
knowledge and influence the speed and direction of his development.

Europe is involved in this immense debate.  It is gambling its very existence as
the cradle of independent civilization.  I hope that this book will bring home the
importance of the stakes, that it will bring the realization that the game is
already lost if the peoples neglect their community of interest as they play.

---

[1] The term was coined by Mr. Denielou, Chairman of UTC of Compiègne, member
of CERD.

[2] Scientific research tends to increase knowledge; it knows what it is asking, but
not the reply.  It does not know what it will find, still less what man will decide
to do with its findings. Hiroshima was contained in particle physics, but no one
could have imagined it at the outset.  Solid state physics has led to microprocessors, a very complex type of artificial nerve cell, endowed with logic and
memory, and fundamental biology is developing an unbelievable capacity for genetic
manipulation.  Only thirty years ago it would have been impossible to believe that
man would discover the keys to such an enormous power to influence nature and
himself.

# ACKNOWLEDGEMENTS

I cannot express my gratitude here to the many who have helped me in preparing this work.  Some of them will recognize the echo of their advice, and many more will smile, I hope, when they see how little their opinions have changed my viewpoint. Be this as it may, their views have given me much food for thought.

My special thanks are due to those who have assisted me within the organization of the Commission of the European Communities, in particular to Dr. G. Schuster, Director General for Research, Science, and Education, and his staff.

Acknowledgement must, of course, be made to the invaluable contribution of my colleagues of the European Committee for Research and Development, without whom the task would not have been undertaken.

Thanks also to the little team meeting at the Fondation Internationale des Sciences Humaines in Paris, and to the Director General, Henri Cavanna.

Dr. Guido Brunner, Member of the Commission, and Professor Ilya Prigogine have kindly written forewords to this book.  May I say how much I was moved by their gestures.

# INTRODUCTION

*Science and the Second Renaissance of Europe* reproduces almost in full a report
which, as Chairman of CERD (European Committee for Research and Development), I
submitted to the Commission of the European Communities. This was under my own
responsibility but with the encouragement of my colleagues, the members of this
advisory body. The report was not intended to tackle the European problem as a
whole; its aim was merely to propose a suitable policy for scientific and technical
research within the Community. However, research paves the way for meeting the
future. Reflection on its aims and its ways and means calls for a good understanding
of the past and the present, combined with a dynamic view of the evolution of
internal forces and a vision of the future of Europe.

Although my colleagues in CERD are not personally committed to any specific opinions
which I express in this work and which will, I am sure, provoke many objections
(a constructive outcome), I believe it fair to say that there was almost complete
unanimity in affirming certain subjects of concern and in asking certain questions.

(a) A profound movement is sweeping through the world of research, an acknowledge-
    ment of the social and moral responsibility of scientists. Scientists cannot
    be disinterested in the use made of their results. They know that they must
    answer to public opinion for the usefulness of their work and that they must
    co-operate in the study of the socio-economic consequences of the technological
    evolution which they originate. This movement is universal. It ignores the
    frontiers of states and communities. It has not yet found the mechanisms
    through which it can act. Its credibility *vis-à-vis* the decision-makers is
    debatable since "scientists" and "professors" are not deemed to be good judges
    of economic, social, or political situations. Contact with the public at large
    is difficult since their language, which bears the stamp of intellectualism, is
    ill-suited. The message transmitted by the mass media is distorted by a taste
    for the sensational or the tragic. This situation must be corrected, and it is
    necessary to justify the confidence in science and technology expressed by
    public opinion[1] but apparently not shared by governments, since only technical

---

[1] An opinion poll conducted by the Commission of the European Communities indicates
that an average 70% of the populations of the nine countries has confidence in the
ability of scientific research to improve living conditions.

progress, which takes account of man in his full dimensions, will enable the eight to twelve billion, which is the estimated levelling-off point for the world population during the next century, to live decently.

(b) A second conclusion has emerged from the Committee's debates.  Europe must prepare itself, starting from today, for a new economic order which will not be based mainly on increased growth in the consumption of material goods. Europe must steer, very positively, in the direction of a quest for economies, especially in the field of energy.  It must manufacture goods which last and which can be repaired, rather than products which rapidly deteriorate.  It must recycle scarce raw materials and protect a particularly fragile ecology.  The members of CERD recognize the dramatic nature of the energy problem and would like to see a "low-energy-consumption society" established.  However, they cannot suggest how this could be done.  Under these circumstances, the spread of nuclear production appears to be inevitable; however, it does not seem that sufficient technical efforts are being made to bring the hazards under control.

(c) A third conclusion of major importance can be drawn from the discussions of CERD.  This is the conviction that although the future is largely unpredictable, the present, to a far greater extent than in past history, shapes the future, and that it is possible to predict the speed of certain economic events fifteen years ahead, barring major catastrophes such as nuclear war.  This conviction has nothing to do with inconsistent futurology.  It is the result of two observations:

   (1) In certain fields, such as that of fossil energy, the near future can be predicted fairly exactly as regards the probability of scarcity and variation in the cost of extraction.  As regards the birth rate, the inertia of trends makes it possible to calculate up to the end of this century, to better than 10%, the size of the population and its age structure, even assuming that, over the next ten years, the populations' intentions regarding fecundity were to change radically.

   (2) As regards conversion of production, it commonly takes from fifteen to twenty-five years to pass through all the stages from the prototype to industrial mass production.  This is the length of time that has to be reckoned with for the exploitation on a significant scale of solar energy or breeder reactors; for automobile traffic employing electric traction to become general; or for recourse to be had to poor ores instead of rich ores.  It takes more than thirty years for a forest to grow and at least fifteen to build a dam on a river in a desert area; to irrigate a large area; to train and settle farmers; start agricultural undertakings and obtain crops.  And what is to be said of town and country planning, the organization of a new diversified industrial complex, an inter-river network of heavy-flow canals?  A lot more additional time will be needed to acquire relevant fundamental knowledge and apply it to social and economic problems.

These factors of inertia in the variation of parameters, this need to take, today, decisions which condition the future, induced CERD to associate itself with recommendations for setting up a European forecasting institute which, according to Professor Dahrendorf's proposal, will throw light on the course to be followed over the next thirty years.[1]

---

[1] A possible programme for this institute has been studied by Lord Kennet (member of CERD) and a working party known as Europe + 30.

Overcoming the loss of confidence affecting society throughout the world while proposing new objectives for consumption, working out a "desirable" model for the near future, and determining research aims for our laboratories are, perhaps, as the objectives of a study, an impossible challenge.  However, it did seem to me that, as far as Europe was concerned, we could make progress in the right direction by seeking answers to the following two questions:

(1) Is Europe's cultural identity, inherited from history and forged by geography, felt with sufficient strength and originality for it to be translated into economic and social objectives capable of inspiring scientific policy?

(2) Do European scientific research workers wish to re-define their objectives and are they aware of their responsibilities *vis-à-vis* the post-industrial society which is in the process of being formed?

The first part of the study, conducted at FISH[1] headquarters by Mr. Henry Cavanna, took the form of a series of inquiries and interviews.

For most of the personalities consulted — historians, philosophers, jurists, and sociologists — the question of Europe's cultural identity was not clear despite, and perhaps because of, the considerable weight of published literature on the subject. Those to whom I talked proclaimed considerable solidarity as Europeans, but immediately dwelt upon the disparities between the Mediterranean and the north, the Catholic countries and those of the Reformation, the regions of Roman-Germanic law and those of common law.  A look at history and culture also shows the arbitrary nature of the Europe of the Nine, too large to be really homogeneous, too small to include all the countries with an eminently European tradition such as those on the Iberian and Scandinavian peninsulas, Switzerland, and Austria, not to mention the countries in central Europe under Soviet influence.  Certain members of the EEC feel more intensely their bonds with the United States or Australia, to whom they consider they gave birth, than their bonds with some of their partners in the Community. Generally speaking, many agree that Europe's diversity, variety of languages, and peculiarities of regional culture are an asset.  However, they never explain how this diversity can become the source of a desire to act with a common purpose.

Nor does the reaction of scientists to their moral responsibilities regarding the impact of their work upon the evolution of Society lend itself to simple analysis, illuminated by the concept of the European Community.  Their feeling of solidarity extends to the entire world, or at all events to a much larger part of the world than Western Europe alone.

As for the fundamentalists, their first concern is the need to increase knowledge. This goal is good in itself and needs no further justification.  However, this desire is accompanied by the need for free communication with other specialists, wherever they may be in the world, providing they are suitably competent.  In this respect, the Community of the Nine does not contribute much.  It is too narrow and, if it wished to intervene, there would be a danger of it acting like a restricting filter.  For the rest, the European Science Foundation, with its sixteen member countries, should settle the problem of a communications infrastructure and meeting place.

As for the leaders of industry, research workers, and engineers engaged in applied research or in the effort to achieve industrialization, their attention is fixed on their objectives.  Very few of them have the time, the inclination, or the com-

---

[1] FISH, International Foundation of Human Sciences, Paris.

petence to ask themselves about the cultural value, and the social consequences of
what they do.  The proposed study is very interesting to them but, save for a few
remarkable exceptions, they do not wish to participate.

And so, instead of easily arriving at a consensus on Europe's cultural personality,
its geographical, historical, and political identity, and its common aspirations
towards certain social objectives which might have been at the root of a common
scientific policy, we found only a sympathetic but sceptical reaction regarding the
project's ambition, more conscious of the opposition and diversity than of the
common good.

At the same time, however, it was confirmed that there is a shared feeling of
anxiety about the future, an inability on the part of the nations, individually,
to acquire the stature necessary to achieve effectiveness.  Indeed Europeans do not
know what Europe is, nor where it is going.  They know the future may give rise to
great difficulties, but they prefer not to think about it.  The breadth of the
dangers disturbs them but the summons to make an effort in order to take charge
of their destiny frightens them.

* * *

In order to proceed further it was necessary to draw up a balance-sheet which, under
such circumstances, was inevitably on the advice of Jean Monnet.  The balance-sheet
is based upon the work of a number of experts and a symposium organized by FISH in
Luxembourg from 3 to 5 October 1976.  The present report is a very personal attempt
to summarize this work.  The summary is not, by any means, by way of being a group
consensus.  In order to provoke reflection and call forth opposition, I have taken
the risk of setting out an argument based on the following viewpoints adopted
*a priori:*

1. <u>A cultural viewpoint.</u>  At the point we have reached in economic development
there is a lag between culture and technological progress.  The power of the tools
of production and its secondary effects triumph over the political will and
cultural aspirations.  This situation must be reversed, and in the cultural
subsystem must be found the reasons for re-orienting the economic subsystem.  In
other words, the human sciences must step in when the knowledge acquired by the
exact sciences is exploited in order to decide the objectives and monitor progress.

2. <u>A viewpoint regarding future prospects.</u>  Questions must be asked of the future
by all available means.  More particularly, despite their obvious flaws and while
keeping an eye on their shortcomings and without ever regarding them as an oracle —
which means that the myth surrounding them must be demolished — recourse should be
had to the principles of systems analysis and applied models[1] for overall social
and economic forecasting.  For the political decision-makers or leaders of industry,
rejection of these means of investigation would be tantamount to committing an
error comparable to that of a doctor, biologist, or economist who refused, in
judging society, to take account of the cultural forces emanating from history,
law, or philosophic thought.

In this study I have been able to use certain results obtained in Cleveland with
the Mesarovic-Pestel model, which was kindly placed at our disposal.

3. <u>A non-Malthusian viewpoint.</u>  Europe cannot satisfy itself with zero growth but
it can no longer, as can the United States or the Soviet Union, count on wealth in
the form of space and primary resources.  If it does not wish to accentuate its

---

[1] See Annex I.

economic dependence to the point at which imbalances on financial accounts would be
unacceptable to its trading partners, it must develop a new type of growth, based
upon the consumption of goods that are economic and not on the waste of raw materials
and energy.  It must increase its capacity for making specific development products
for the countries in the process of industrialization, by organizing with them
reciprocal technical and commercial relations.

4. **A neutral viewpoint with regard to different political profiles.**  The study
leads inevitably to reflections of a political nature.  The reader will perhaps be
surprised not to find herein any choice in favour of one doctrine or one ideology.
This is not a refusal to be committed but a resolution to stick to the facts
without associating myself with any of the major options of the nineteenth century.
These I consider, for the most part, to be empty of meaning, faded, word-worn, and
distorted by the image of the *déjà vu*.  In addition, these options are profoundly
ill-suited to the realities of the modern world.  Instead of being based on
ideologies, the study systematically endeavours to observe reality, the interaction
of cause and effect, by using the new ways of thought offered by systems analysis.
The reader is then free to introduce his own political options to embellish the
solutions he favours to the problems posed.

*   *   *

We have now embarked on a search to discover what the European identity can be,
accompanied by meditation on its long-term consequences.  In a preliminary analysis
I have tried to draw up a balance-sheet that is as objective and as uncontroversial
as possible.  My thinking becomes more deliberately personal when I come to the
aspirations and projects that Europe appears to have in its womb.  However, this
projection to the future is by no means, to my mind, an idealistic view of what the
European society of tomorrow ought to be.  Rather, it is a view of what tomorrow's
European society is, to some extent, predestined to become on account of the
endogenous and exogenous forces arising from its environment.  As will be seen,
the forces of the environment are assumed to be an emanation from the past, of
historical and cultural origin; an emanation from geography, the birth rate,
economic and political competition, and the economic needs of the rest of the
world.  There is no question of pronouncing a moral judgement on tomorrow's society.
Even less is it a question of assuming that political action could prefabricate the
future.  The future, in essence, remains unknown to us.  No man has the right to
predetermine it for his fellows.  The future must remain free but have light thrown
on it, in so far as it is predictable, in order to eliminate or mitigate the major
threats.

What can be said now about Europe is that if it is decided to do nothing or to do
little, a crucial decision regarding its future will have been taken and a
rendezvous with tragedy will have been made as regards economic dependence,
unemployment and social tension.

*   *   *

Fundamentally, this study has a utilitarian aim: to promote the formulation of a
Community scientific and technical research policy.  The main aim of this policy
would be to enlarge the freedom of choice open to the generation of decision-
makers thirty years ahead.  If results of certain studies are available,
governments will be able to make their choices in full knowledge of the facts.
If the effort to acquire these results is not made, either the solutions will be
imposed or decisions will be taken blindly

The reader will probably regret that the conclusions stop short at the consideration
of research policy.  As far as European economic forces are concerned, the weak

point is chiefly the lack of an industrial and economic policy.  In the rooms of
the House of Europe set aside for innovation, the fires do not always draw well;
instead of helping to warm the partners, they sometimes tend to smoke them out.
If this situation is prolonged it will have disastrous consequences in other fields
that will adversely affect commercial competitiveness, the balance of payments,
and the level of employment.

Europe must be taken as it is today, incapable of political unification in the
conventional meaning of the term.  But the true problem in the Community is perhaps
not to install a hierarchical power superior to the national powers.  Europeans are
too attached to their diversity to allow this, and perhaps, all things considered,
it is better thus.  It is necessary to invent new systems of intervention that will
interconnect decisions at national and Community levels, so as to generate
movements of solidarity or to allow all to benefit from scale effects of European
dimension.  This will not make the Community a super-government but an instrument
for the promotion of complementary or supplementary activities that would not have
been originated, and would not have prospered, in its absence.  It will be multi-
lateral.  It will allow emulation between the partners, which has up to now been
one of the sources of European progress.  And the problem will be solely to foster
the synergy of the specifically Community activities with national activities.
The implementation of a research and development policy by the Commission of the
European Communities could provide an opportunity for trying out these new
intervention arrangements; it would be a kind of trial on a reduced scale of what
new types of Community policies could become in more sensitive and more ambitious
fields.  This is the background against which the proposed solutions have been
devised.  The advocates of a highly unified Europe may consider them merely as one
stage towards more formal structures; those who are mainly concerned to preserve
national identities will be reassured and able to give their support to the
experiment without misgivings.

Nothing will be done for Europe, however, until progress is made towards the
"addition from the soul" to the European effort in response to the invitation in
the Tindemans Report.  I hope that the reader will find in my work incentive
towards a turning point to be negotiated by our society.  The interest of the text
lies not in the quality of the guidelines proposed, but in the encouragement to
new thinking that it may provide.

Of course, Europe cannot negotiate this turning point on its own; it must become
complementary with the other major geographical areas.  If it wished to isolate
itself, the forces of interdependence which dominate today's world would soon
recall it to harsh reality.  But Europe should claim a vocation: to become one of
the "social development laboratories" which humanity needs so greatly today.  It
has yet to realize, by proposing a project, that it could play the role; after
that, it is probable that Europe will decide how it is to be done.

Part I
# THE SEARCH FOR THE EUROPEAN IDENTITY

Chapter 1

# THE BALANCE-SHEET

As opposed to the remainder of this study, which may in some respects be regarded as a plea on behalf of certain arguments, the balance-sheet given below endeavours only to relate facts and comment upon certain estimates resulting from the extrapolation of trends noted over recent years.

The figures and the curves quoted are subject to errors.  The statistical apparatus at our disposal for assessing the present, and even more so the past, is unsatisfactory.  Projections into the future are the result of debatable hypotheses.  If, however, like doctors, we devote our attention to orders of magnitude in phenomena while neglecting detail, the conclusions are firm and make meaningful diagnosis possible.

This balance-sheet shows Europe as it appears today in the light of the political situation, demography, and geographical and economic data.

## THE POLITICAL EVOLUTION

Without going back into very ancient times, which continue to exercise an influence although almost exclusively in the cultural field, the history of the last few centuries is rich with lessons which help us to understand the importance of the change which has been imposed upon Europe as regards political and military power.[1]

For four centuries Europe engaged in conquering and then occupying space on this planet, creating an immense empire which was at its apparent peak on the eve of the Second World War.  With the exception of China, Japan, and a number of states in Asia, which were never deeply penetrated, all the continents have been subject to this expansion, some to settlement and others to exploitation by European colonialism.

The first phase of expansion was essentially Latin.  It was in the direction of the West Indies and was undertaken by the Iberian powers with the support of maritime techniques which had come from Italy.  It extended from the end of the fifteenth

---

[1] This chapter is based on a bibliographic survey carried out by Mrs. A. Devred on behalf of FISH (February 1977).

to the beginning of the nineteenth century, when Spain and Portugal lost control of their American possessions.  However, they left behind them their languages, their religion, a marked cultural imprint, and a large proportion of their ethnic elements.  Latin America is the daughter of Iberia.

The second phase of expansion, slowly prepared in the fifteenth and sixteenth centuries, was by France, Great Britain, and Holland during the seventeenth and the beginning of the eighteenth century.  Maritime superiority was lost by the Spanish and Portuguese to the hands of the north-west Europeans who were more enterprising and concerned with discovering new territories to populate.  The Caribbean islands changed hands.  France and Britain competed for the North American continent.  Trading centres, which constituted the bridgeheads for colonization, were founded in India and Indonesia.  Australia was discovered and became New Holland.  France claimed rights to Madagascar.  Africa had as yet but little to say except as a supplier of slaves.

On the North American continent the adventure started with the independence of the United States.  In that country the aboriginals living there before conquest have practically disappeared, so that the ethnic elements are almost exclusively European with the exception of black forced immigrants.  The languages of Western Europe, its culture, its different forms of Christianity and especially Puritanism, its enterprising spirit, its capacity for invention and for giving expression to democratic forms of political power, helped in the explosive rise to power of a new nation to which was open the conquest of virgin space rich in possibilities for agricultural, mining, and finally industrial development.  Right through its period of growth the United States continued to swell its population by immigrants from northern and Central Europe and the Mediterranean.  The North American melting pot gave them homogeneity.  The United States is a branch of Europe transplanted into a basically different environment in terms of available space, the scale of natural phenomena such as rivers, deserts, and climate, and the resources in the soil and the subsoil.  When US supremacy over Europe has affirmed itself, the US people may be regarded by the Europeans themselves and by other peoples as "the Europeans who succeeded".

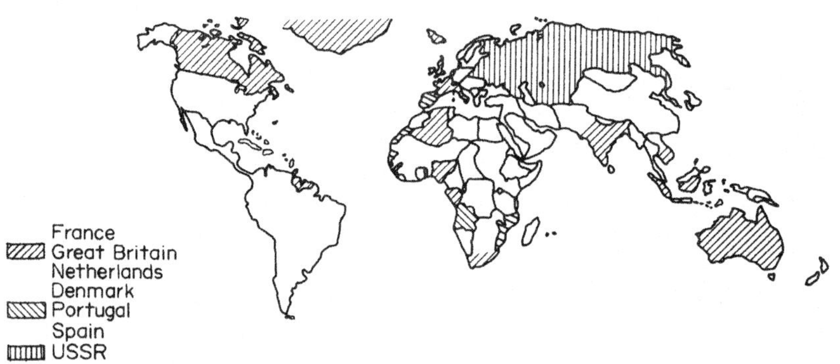


Fig 1. European colonial empire in 1870.

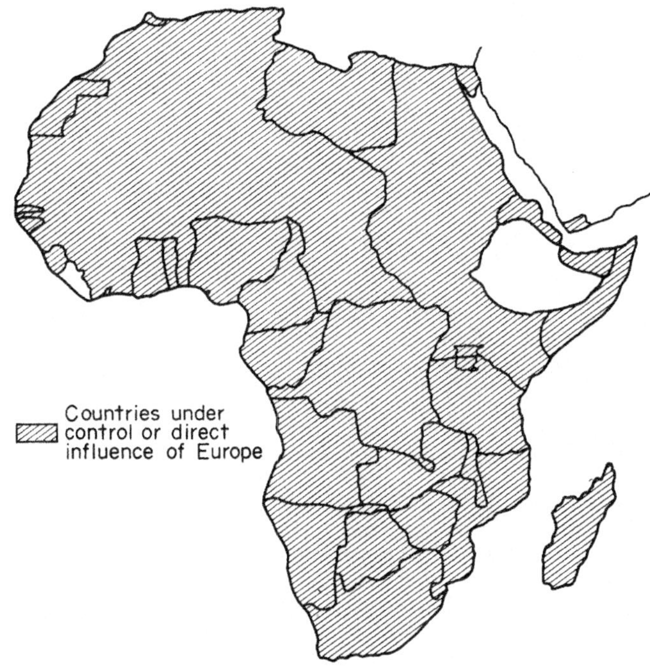



Fig. 2. Africa in 1914.

After the Napoleonic wars, Europe bound its wounds and devoted its strength to
promoting the Industrial Revolution, the main scene of which was the regions in
the north and the north-west.  The expansionary phases calmed down over almost
three-quarters of a century, and then flared up again even more vigorously after
1870 (Fig. 1).  By then Africa had been almost entirely conquered and shared out
(Fig. 2).  With the exception of Thailand, southern and South-east Asia was
subjected in the form of colonies or protectorates.  The collapse of Turkey in
1918 opened up to the allied powers the road to the Middle East and the whole of
North Africa (Fig. 3).  On the eve of the Second World War, Italy conquered
Ethiopia, Africa's last independent nation.  Australia, Canada, and South Africa
were alongside Great Britain in a Commonwealth covering five continents.

European domination was then apparently at its peak.  Without counting Russia who,
during this period, confirmed its hold over Siberia while ceding Alaska to the
United States, the European colonial powers, whose metropolitan base was not
more than 1.8 million square kilometres, associated within their political system
and placed under their military control more than 50 million square kilometres of
overseas possessions, i.e. more than one-third of the world's land surface.

One-quarter of the world population was under the direct domination of Europe's
political power.  Control of the seas was disputed only by the United States.  The
sources of enormous quantities of raw materials, both mineral and vegetable, were
controlled by the states of Europe or by the power of the capital of Europe's
entrepreneurs.  Middle East oil was subject to the political influence of Great
Britain, whose currency possessed international value.

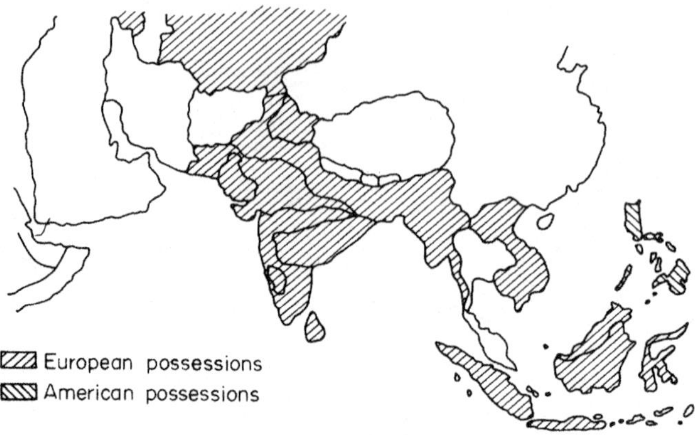


Fig. 3. Southern Asia in 1914.

However, Europe was profoundly divided, and the Second World War came to an end
amid the Continent's ruins.  The leading roles fell to the United States and the
Soviet Union.  The vicissitudes of the war had given the dominated peoples a taste of
independence and the idea that it might be achieved.  The moral force which argued
that colonization was good because it brought civilization and technical progress
to peoples who did not possess it was entirely reversed.  World opinion, amplified
by the United Nations, was in favour of independence for the former colonies.  The
Bandung Conference in 1955 proclaimed the will of the peoples of the Third World.

In 1956, Western Europe stood by without reacting in the face of the vicissitudes
of the Hungarian insurrection.  Great Britain and France yielded at Suez before
the combined intimidation of the Americans and the Soviets.  In doing this they
announced to the rest of the world that they no longer possessed their own
political power.

From then on, the retreat from what was left of positions acquired over four
centuries was certain (Fig. 4).  It was to be accomplished inexorably, and not
without some shudders, over a period of less than twenty years, terminating in
1975 with the independence of the Portuguese territories in Africa.  Meanwhile,
Western Europe became fully aware again; the freedom of peoples to dispose of their
own destiny is, indeed, a cultural product which came from Western Europe itself.
It is consistent with the practical and moral ideas regarding democracy and
responsibility which the Occident considers to be essential values.  In addition,
the idea spread that economic wealth and standard of living are not a function of
the availability of basic products but of the efficiency of systems of production
and distribution.  Freed from the burden of its colonies, Europe became free to
devote its attention to domestic progress and an export effort.

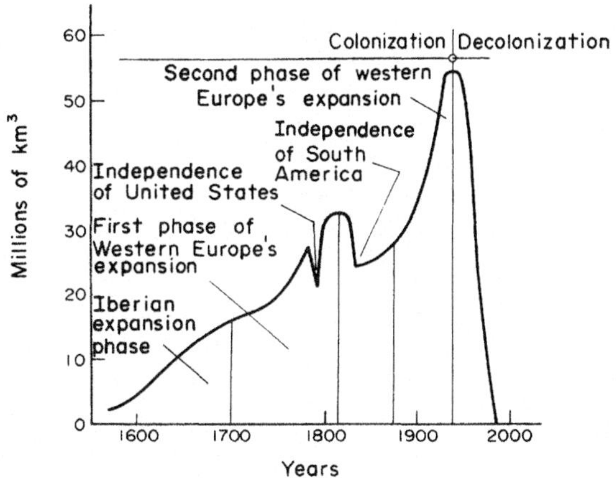


Fig. 4. Rate of variations in area of European colonies or
protectorates from 1600 to the present day.

On 15 August 1971, without consulting its European partners, the United States took
capital decisions in monetary matters and affirmed the absolute supremacy of the
dollar; Europe's reaction was zero or to no avail. In the autumn of 1973 the oil
producers agreed to quadruple the price of a fuel which accounts for 57% of the
energy resources of the European Economic Community. Instead of the reaction being
one of uniting to establish targets for economy and develop new sources which
became dramatically necessary, national egoism enjoyed almost full play. The
Middle East powder barrel is saturated with exports of arms employed as an
instrument for balancing budgets. It is, doubtless, useless to dwell here on the
weakness of Western Europe's military positions; its naval forces are no longer
master even of the Mediterranean. As far as the land and air forces are concerned,
they are entirely dependent upon US goodwill (Table 1).

TABLE 1. Armaments table[1]

| Country | Forces | Expenditure (million dollars) | No. of tanks | No. of fighter aircraft | No. of large surface vessels | No. of conventional submarines |
|---|---|---|---|---|---|---|
| France | 527 000 | 9 400 | 2 070 | 461 | 63 | 19 |
| US | 2 130 000 | 92 800 | 10 530 | 6 900 | 179 | 75 |
| USSR | 3 575 000 | 103 800 | 40 000 | 6 065 | 236 | 265 |
| Great Brit. | 345 100 | 10 380 | 1 080 | 530 | 77 | 28 |
| Italy | 421 000 | 4 220 | 1 300 | 372 | 30 | 10 |
| FRG | 495 000 | 16 260 | 3 700 | 585 | 22 | 24 |
| China | 3 250 000 | n.c. | 8 500 | 4 400 | 31 | 52 |
| Egypt | 322 500 | 6 103 | 1 975 | 608 | 8 | 12 |

In the case of France, the numbers do not include the gendarmerie and paramilitary
forces.
Source: Military Balance (1975-1976).

---

[1] Taken from *France-Forum* No. 149, July/August 1976, Dominique Kergall
"Quel avenir pour l'Occident?"

The events of recent years have thus demonstrated that politically Europe is
extremely weak.  For the first time in their history, Europeans are obliged to
observe the wisdom of ceasing to wage war in order to dissipate a surplus of
aggressiveness.  Europe, if it wishes to defend its identity, can take the
offensive only on the basis of its cultural strength and its economy's potential
for competition.

This situation is not yet generally accepted, however.  Most Europeans have no
idea of the danger to which this political weakness exposes them.  And their
partners in the developing world, newly independent, are only slowly developing their
own identity, with the impression that they are still subject to an economic
dependence that weighs heavily on them.

On the other hand, Europe's military and political weakness should make it easier
for the peoples of the developing countries to regard Europeans as partners having
absolutely no imperialist intentions (or at least power) who are naturally
complementary on a basis of reciprocity (Table 2).

### TABLE 2. The main stages of decolonization

| | |
|---|---|
| 1945 | Proclamation of the independence of Indonesia |
| 1946 | Creation of the Union Française |
| 1947 | Independence of India |
| 1948 | Independence of Ceylon and Burma |
| 1951 | Independence of Libya |
| 1954 | Geneva agreements on the Indochinese peninsula |
| 1955 | Independence of Morocco/Bandung conference |
| 1956 | Independence of Tunisia, Egypt, Sudan, and Malaysia |
| 1957 | Emancipation of Ghana |
| 1959 | Dissolution of the Union Française and establishment of the Communauté Française |
| 1960 | Independence of the territories of West and Equatorial Africa Birth of the Republic of Somalia Independence of Zaire |
| 1962 | Independence of Algeria |
| 1963 | Independence of Zanzibar, Kenya, Malawi, and Zambia Emancipation of Malta |
| 1975 | Emancipation of Angola and Mozambique |
| 1976 | Emancipation of Western Sahara |

### POPULATION TRENDS

The size of the population in absolute terms and in relation to the remainder of
the world, its density in the space available for population, and its composition
per age category are fundamental data if it is desired to comprehend certain
economic and social phenomena.  In this field, Western Europe's situation is, by
comparison with the rest of the world, in a phase of complete change.

For a long time Western Europe was the world's major reservoir of men.  Upon
emerging from the Middle Ages it was, after China and India, the most populated
region.  Although its density of population per square kilometre never reached
the levels observed today, it has always been Europe's endeavour to find
territories to populate outside itself.  This desire partly explains its efforts
to achieve overseas conquests.

Europe's fecundity has created considerable migration flows which have supplied the foundation for populating the Americas, South Africa, and Australia.  However, despite having ceded part of its population to other continents, Europe maintained until the nineteenth century its position as a very densely populated territory by comparison with the rest of the world.

The trend of events was reversed in the nineteenth century.  Western Europe's relative share in the world population fell progressively until the watershed of the fifties, when the population decrease speeded up dramatically.  Reduced fecundity on the part of the Europeans contrasted with a sharp population growth in the developing countries, where the fight against perinatal mortality was making spectacular progress.

It is important to take recent events into account and even more those which can be forecast with a fairly safe margin of approximation for the next twenty-five years. Although modern contraceptive techniques and legalized abortion can today profoundly change the trends observed in the past as regards fecundity, there are in fact sufficient inertia factors for relatively short-term population forecasts to be valid.

For the past, United Nations statistics have been taken as a basis.  For the period 1975-2050 the figures are from the Mesarovic-Pestel model.[1]  Questions were fed into the model on the basis of two extreme hypotheses, one according to which the fertility trends observed during the years 1950-1970 remain unchanged, and another whereby a very strict anti-birth policy is enforced immediately in the developing countries. The truth lies between these two hypotheses.  As far as Europe's future place is concerned, the degree of uncertainty is of no importance.  Whatever happens, its place must be quantitatively minute; it will count only qualitatively.  This is the fundamental lesson to be learnt from these forecasts.  As far as the nations which make up the Europe of the Nine are concerned, in the most populated among them — Great Britain, Federal Germany, Italy, and France, which each accounted in 1950 for more than 2% of the world population — children born today will see this percentage fall rapidly to less than 0.6%.  This is unless the trends observed in most recent years in the direction of a further fall in the birth rate are confirmed. The level of the population is maintained at a rate of about 2.1 children per woman.  In Europe, particularly northern Europe, rates of 1.8 or lower have recently been observed.  If the phenomenon continues, a country such as Federal Germany would fall below 0.4% of the world population in 2050 — even if the developing countries were to reduce their birth rate considerably (Figs. 5 and 6).

And so the EEC, while ceasing to become a territory from which there is emigration, is tending, despite an immigration flow in the opposite direction, to occupy in less than twenty-five years a humble place in the world, in terms of numbers of human beings, more humble than the relative position occupied in the eighteenth century by any one of the great nations now a member of it (Table 3).

---

[1] The results obtained with the help of the Mesarovic-Pestel model were reviewed by INED specialists for the period 1975-2000.  They were sufficiently near the United Nations most recent forecasts for there to be no need to adjust them.  The figures given in the tables for the Europe of the Nine come from INED (Institut National d'Etudes Démographiques, Paris).

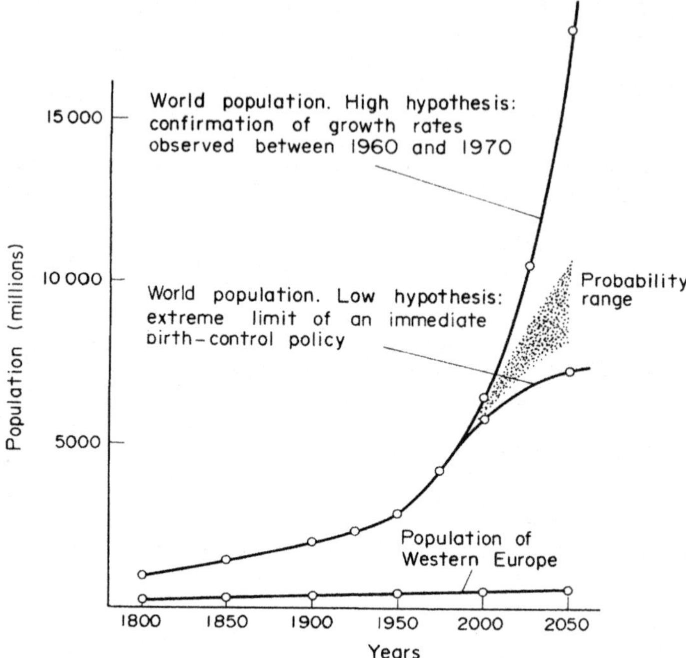


Fig. 5. Forecast of comparative trends in the world and
Western European populations.

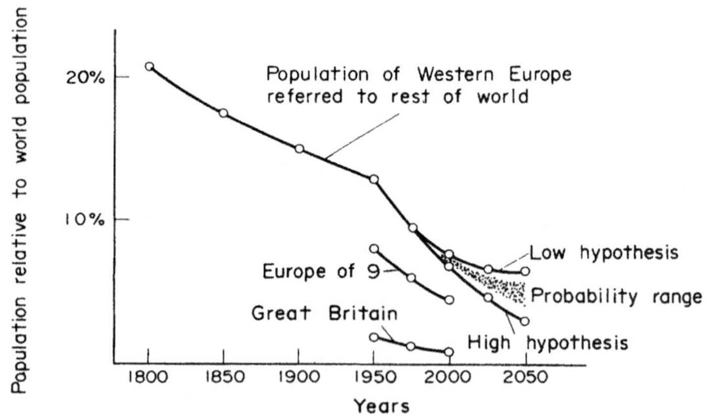


Fig. 6. Forecast trend in the ratio of European to world population.
Western Europe:  Mesarovic-Pestel model.
Europe of the Nine: INED.

TABLE 3. Forecasts of population trends (millions)

| | World Population | | Western Europe Population[a] | | Europe of the Nine Population[b] | Percentage Western Europe/World | |
|---|---|---|---|---|---|---|---|
| | Low hypothesis | High hypothesis | Low hypothesis | High hypothesis | | | |
| 1800 | 960 | | 200 | | | 20.6 | |
| 1850 | 370 | | 242 | | | 17.5 | |
| 1900 | 1 950 | | 293 | | | 15.0 | |
| 1925 | 2 300 | | 320 | | | 13.9 | |
| 1950 | 2 700 | | 351 | | 215 | 12.9 | |
| 1975 | 4 120 | | 404 | | 259 | 9.8 | |
| 2000 | 5 800 | 6 400 | 440 | 450 | 260 | 7,6 | 7 |
| 2025 | 6 800 | 10 400 | 460 | 490 | | 6,7 | 4,7 |
| 2050 | 7 100 | 17 700 | 465 | 540 | | 5 | 3 |

Note: Western Europe is here Region No. 2 in the Mesarovic-Pestel model.
    This includes not only the Nine but also all non Soviet-bloc
    European countries.
[a] Mesarovic-Pestel model figures.
[b] INED figures.


This situation must be considered highly probable in the absence of catastrophes.
Contesting the figures — attributing an error of approximately 20%, which is
certainly excessive for the period separating us from the year 2000 — does not
change the orders of magnitude at all, any more than would be the case if an
extremely determined policy for encouraging the birth rate were pursued in
Europe alone.

Western Europe's comparative demographic shrinkage by comparison with the
remainder of the world is not an isolated phenomenon. It is found in all the
industrially developed regions in comparison with the developing countries. It
affects both the countries with an economy based on free competition and those
with a planned economy. If, in outline and omitting Japan, we regroup the
industrialized countries into three main regions — the North American continent,
the Soviet bloc, and Western Europe — it will be seen that in the years to come
there will build up at the frontiers of these three areas considerable demographic pressures derived from Latin America, South-eastern Asia, and the Arab
world respectively.

A comparison of the population trends of Western Europe and the Arab world is
particularly significant on account of the shared responsibility for the
Mediterranean, the economies' complementarity and the proximity of their
cultural heritage. Figure 7, provided by the Mesarovic-Pestel model, shows

that with effect from the year 2000, i.e. almost tomorrow, young Arabs[1] below the
age of 15 will be more numerous that young Western Europeans,[1] whereas in 1950 the
percentage for young Arabs was scarcely 40%.

The structure of the populations by age group will, of course, be radically
different.  The opposing partners in the situation will be an Arab world composed
mainly of young persons and a European community peopled by persons of a ripe age,
not to say old.  This is the second inexorable forecast which can be made regarding
the population (Figs. 8 and 9).

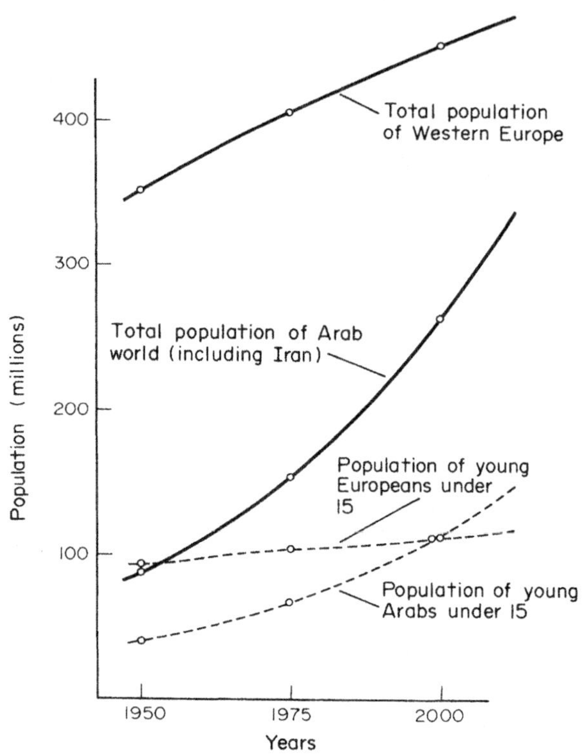


Fig. 7. Comparative trend for the populations of Western Europe
        and the Arab countries including Iran.  Total populations
        and young persons under 15.  Hypothesis: continuation of
        current trends.

---

[1] It should be remembered that in the Mesarovic-Pestel model the Arab countries
include all the countries in North, West, and East Africa, the Middle-East
countries and Iran.  Western Europe covers all the countries not integrated into
the Soviet bloc, more particularly Yugoslavia, Greece, and Turkey.

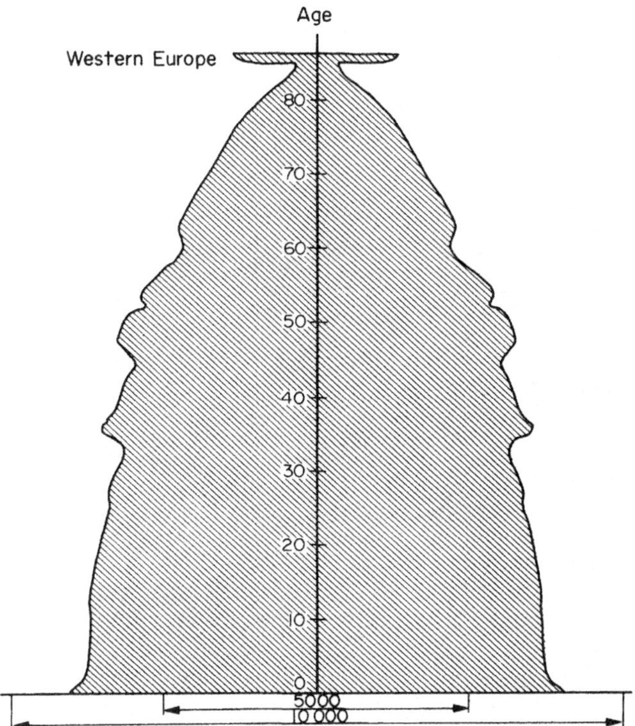


Fig. 8. Study of age structures: forecast for
Western Europe for the year 2000.

## GEOGRAPHICAL DATA[1]

Western Europe constitutes a geographical whole which is remarkably homogeneous by
reason of its temperate climate, even in the most northern regions, which are
under the influence of warm sea currents, its irregular coastline, its orography and
hydrography where extremes are absent, and the ease with which its territory can be
penetrated by man for the purpose of transport, town and country planning, agriculture,
and industrial development.

All Europe's dimensions are on a scale appropriate to agricultural, artisanal, or
intellectual man.  No natural phenomenon is excessive.  The lines of communication
are short.  The largest river, the Rhine, is a dwarf by comparison with the Amazon,
the Mississippi, the Nile, the Yang-Tse or even the Volga.  The Alps, dissected by
morainic valleys, are far more open to human traffic than is the Andean Cordillera
or the Himalayas.  There is neither desert nor jungle.  The land has been cleared
of all species of animals hostile to man.  The permeation of nature by man, and a
population density that has been high for centuries has meant that there now exist
only flora and fauna which are tolerated and useful to human life.

---

[1] This chapter draws strongly on the paper read by Mr. Kormoss at the symposium
organised by FISH in Luxembourg in October 1976.  See proceedings: "Some aspects of
the geographical and demographic bases of European integration".

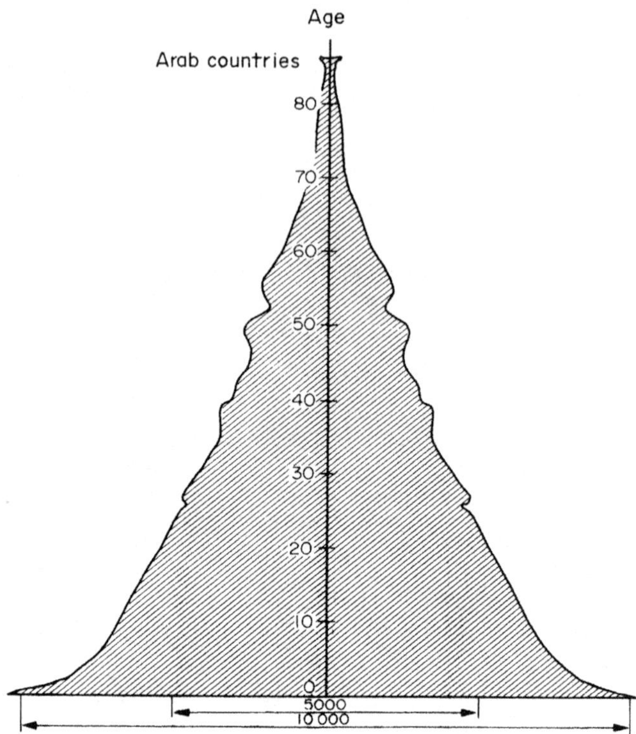



Fig. 9. Study of age structures: forecast for the
Arab countries for the year 2000.

Western Europe lacks room. On average it is the region in the world where the
population is most dense (Figs. 11 and 12). In this respect, among the
industrialized countries, only Japan experiences similar problems which there are
rendered even more acute on account of the small area of land suitable for
settlement (approximately 16% as opposed to 30% for Europe).

Western Europe is relatively poor in easily accessible rich ores. Intensive mining
of the most readily available resources has gone on throughout history with the
result that the most accessible deposits are exhausted. What is true of gold,
diamonds, tin, lead, mercury, and coal is true also of the more recently exploited
ores such as bauxite. In addition, sources of fossil energy are scarce or
difficult to reach (ocean deposits). Europe's hydrology is not very conducive to
the production of large quantities of energy. If Europe wants to obtain the raw
materials it needs from its own soil, it will on average have to spend more than
most of the other major industrialized countries, work less economically, and use
more energy for the processing of low-grade ores.

The scenery is not conducive. The littoral opens on to shallow seas which are
closed or almost closed. The rivers flow through highly urbanized areas. It is
intolerable that they should be used as carriers of waste. The large forest areas
are not much developed.

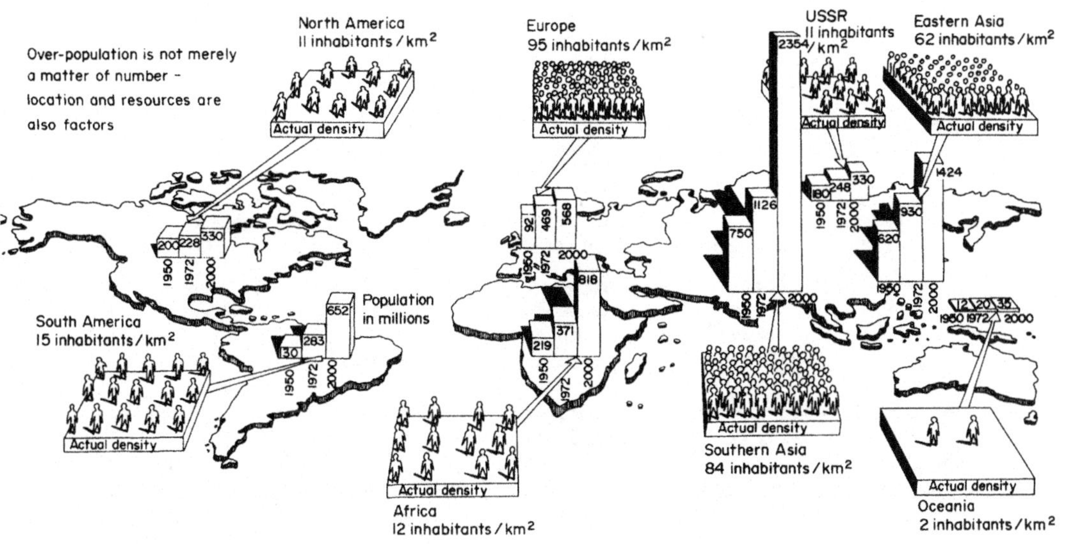


| Country | Population (millions) | Density (inhabitants/km²) |
|---|---|---|
| China | 800 | 83 |
| India | 576 | 172 |
| USSR | 248 | 11 |
| USA | 208 | 22 |
| Indonesia | 126 | 78 |
| Japan | 107 | 284 |
| Brazil | 101 | 12 |
| Bangladesh | 72 | 524 |
| Pakistan | 64 | 74 |
| Federal Republic of Germany | 61 | 248 |
| Nigeria | 58 | 70 |
| United Kingdom | 55 | 229 |
| Italy | 54 | 180 |
| Mexico | 54 | 25 |
| France | 51 | 95 |
| Netherlands | 13 | 380 |
| Canada | 21 | 2 |
| Libya | 2 | 1 |

Fig. 10.  With 800 million inhabitants, China is four times less populated than the Federal Republic of Germany, and India two times less populated than The Netherlands. It is the Europeans who risk lack of room — the others, lack of resources.

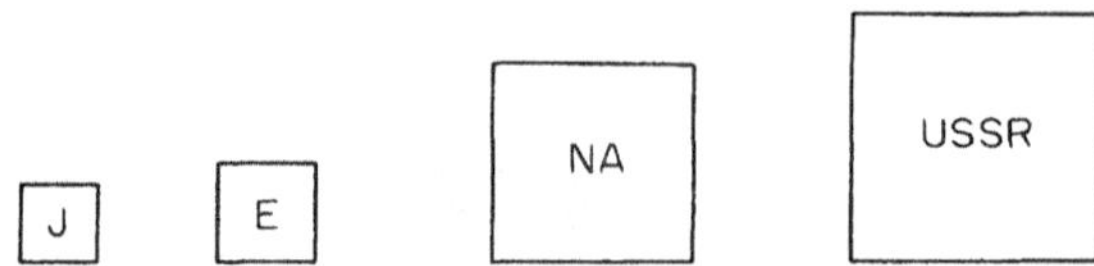



Fig. 11. Space available to the inhabitants of the main
developed regions:[1]

for each person in Japan: 0.35 ha
for each person in the EEC: 0.60 ha
for each person in North America: 4.70 ha
for each person in the USSR: 9.00 ha

Europe's natural dimensions are not favourable to mammoth industry.  When new
technologies have to be tried out in wide open spaces, as is the case of military
research that can have important civil spin-off, Europe is handicapped.  The same
is true when harmful waste needs to be recycled.  Europe lacks vast continental
rivers easily capable of providing cooling or process waters, hugh open spaces where
air pollution is of no account, and open country with a low population which can be
despoiled by mining without affecting man.

In Europe the price of land is a painful economic fact; it is difficult to set up
a chemical plant, oil refinery or nuclear power station close to built-up areas.
Cooling water is scarce — air and river pollution is intolerable — and the transport o
goods on main routes aggravates traffic congestion.  All these factors increase
industrial overheads or necessitate local authority expenditure.

The European scale of nature and the high population density appear to militate in
favour of small production units, which are in any case more acceptable from the
human aspect.  Europe, geophysically and culturally, generally does not welcome
mammoth industrial units.  With mass production costs as they are today, this is
an additional difficulty to be taken into account.


### ECONOMIC EQUILIBRIA

Throughout history man has suffered greatly from insufficient economic capacity,
lack of food, inadequate housing, poor defence against the vicissitudes of climate,
and insufficient educational facilities.  However, through the extent of its
agricultural and artisan activities and industrial development, and through the
advantages which it has obtained from its political power, Europe has experienced
over the last few decades situations which were generally favourable.  This has not
prepared its peoples for frugality.

---

[1] If the United States had the same population density as Western Europe, it would
have 2000 million inhabitants.

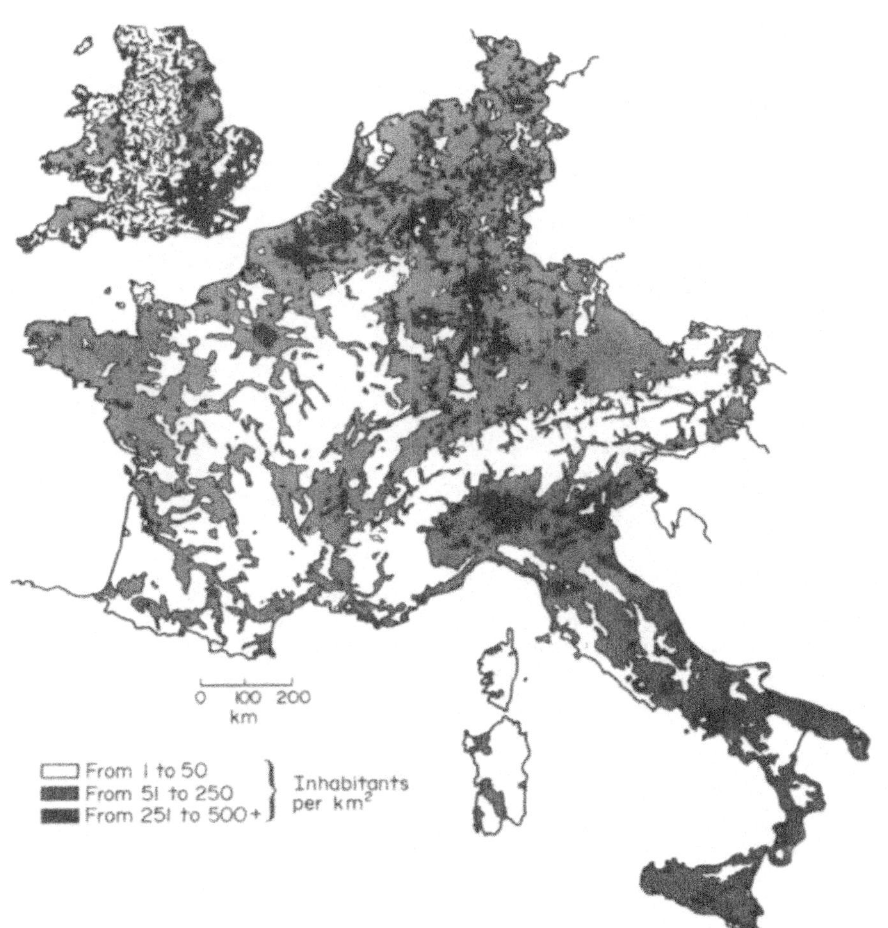



Fig. 12. Population density of the countries of central
western Europe.  Map compiled from the work of
Professor I.B.F. Kormoss, Collège d'Europe, Bruges.

Once the pieces had been picked up after the Second World War,[1] the capacity of its
peoples, mobilized for industrial growth, enabled Western Europe to hoist itself
again to the level of the most-favoured nations as regards potential for the
consumption of all kinds of products.  During the fifties and halfway through the
decade thereafter the countries of Western Europe had as their real industrial
competitors only the United States — itself in full expansion.  Imported products,
raw materials and energy were bought cheaply.  Competition from Japan and South-East
Asia was only beginning.

---

[1] Largely thanks to the Marshall Plan

The horizon rapidly darkened at the beginning of the sixties.  Inflation and
monetary disorders accompanied a break in the rate of growth in most of the
industrialized countries.  With effect from the autumn of 1973, Europe learned
that it was going to be dictated to by certain suppliers such as those of oil and
gas, and perhaps phosphates and other raw materials.  In recent years, the Eastern
nations re-started, under the spur of development needs, their commercial offensive,
which involved a great variety of products such as steel, cars, ships, clocks and
watches, radio and television receivers, electronic components and micro-processors,
textiles, etc.  Japan did this on the basis of its own independence.  South Korea,
Taiwan, Hong Kong, Singapore, and recently-independent countries, such as Indonesia,
were assisted with foreign aid.

For the needs of this study the Mesarovic-Pestel (M.P.) general macroeconomic model
has been used to produce forecasts regarding the deficits in food, energy, and non-
ferrous metals for Western Europe for the period 1975-2025.  The results exhibit
numerous uncertainties:  they cannot take into account the unforeseeable, the
unexpected changes that may occur in the political or economic order or as a result
of the widespread application of the new technologies now being spawned in our
laboratories.  Fifty years ago, no one could seriously have predicted the
consequences of biological progress on population trends nor the effects of nuclear
science, electronics, and computers on our society.  It is logical to believe that
new surprises await us in the technological field in the next thirty years, and it
is by no means sure that we will be any better able to control their effects than
we have been in the past.

Because of these inadequacies in the forecasting models, we give in the Annex the
results obtained from the experiments with the M.P. model.  The reader will find
ample food for thought.  Despite inevitable approximations, the results do at least
indicate orders of magnitude.  They show that the current deficits will steadily
increase in the future in so far as scenarios can be drawn by extending statistical
observations made in the recent past.

To measure the current degree of economic dependence of Europe of the Nine, it
appeared more convincing to portray the situation that emerges from the statistics
put out by the Directorate-General for External Affairs of the Commission of the
European Communities, in the publication *Community trade in 1975: changes as
compared with 1974.*

Table 4 shows Community trade with the rest of the world, quantifying the deficits
in food, raw materials, fuel products, and primary products.  Ratio ranges vary
between 12 and 40%.  With all their talk of milk and butter surpluses, the
Europeans have perhaps forgotten their overall food deficit.  Far more alarming
and worrying, however, are the energy and raw material deficits, as the volumes
involved are considerable.

Figures 13 and 14, from the same document, show perhaps even more clearly the very
special balances in the Community economy.  They indicate that Europe serves as a
sort of vast factory which makes chemicals, manufactured products, and capital
goods (85% of its exports) for other countries, from which it purchases most of its
raw materials, its energy, and much of its food requirements (58% of its imports).

The unbalanced shares of the various member countries in the Community's trade with
the rest of the world, one of the features of the current situation, add up to a
general dependence on imports of primary products, as Table 5 shows (same source).

TABLE 4. Community trade with the rest of the world in 1975
broken down by product
(million EUR and %)[a]

| CST | Products | Imports | | | Exports | | | Trade balance | Export/Import ratio |
|---|---|---|---|---|---|---|---|---|---|
| | | Mil. EUR | 1974 = 100 | % | Mil. EUR | 1974 = 100 | % | (Mil. EUR) | (%) |
| 0-9 | Total external trade (EUR-9) | 117 357 | 94 | 100 | 113 555 | 105 | 100 | -3 802 | 97 |
| 0 | Food | 14 755 | 106 | 13 | 5 855 | 101 | 5 | -8 900 | 40 |
| 1 | Beverages and tobacco | 1 215 | 109 | 1 | 1 582 | 107 | 1 | +367 | 130 |
| 2 | Crude materials | 15 367 | 82 | 13 | 2 111 | 84 | 2 | -13 256 | 14 |
| 3 | Fuel products | 36 368 | 90 | 31 | 4 451 | 95 | 4 | -31 917 | 12 |
| 4 | Vegetable oils and fats | 782 | 50 | 1 | 450 | 100 | 0,4 | -332 | 58 |
| 5 | Chemicals | 5 055 | 85 | 4 | 12 843 | 88 | 11 | +7 788 | 254 |
| 6 | Manufactured goods classified by material | 17 033 | 87 | 15 | 24 824 | 94 | 22 | +7 790 | 146 |
| 7 | Machinery and transport equipment | 15 635 | 107 | 13 | 50 104 | 120 | 44 | +34 469 | 320 |
| 8 | Miscellaneous manufactured articles | 8 213 | 109 | 7 | 9 271 | 103 | 8 | +1 058 | 113 |
| 9 | Goods and transactions not classified elswhere | 2 934 | 193 | 2 | 2 064 | 123 | 2 | -870 | 70 |
| 0-4 | Primary products | 68 487 | 91 | 59 | 14 499 | 96 | 13 | -54 038 | 21 |
| 5-8 | Industrial products | 45 936 | 97 | 39 | 87 042 | 106 | 85 | +51 105 | 211 |

[a] EUR: European Monetary Unit = 0.888 670 8 grams of gold = US$1.32 (in 1975).

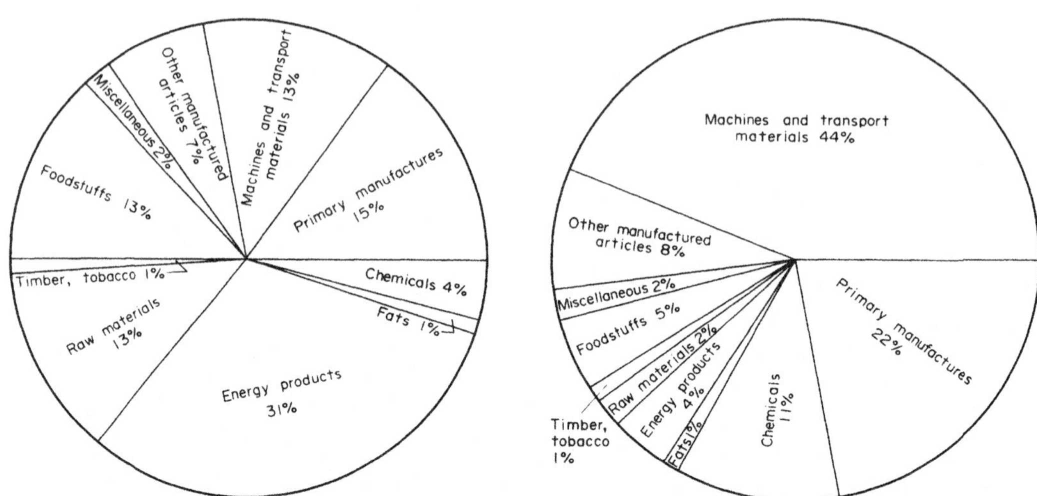


Structure of the Community's trade with the rest
of the world in 1975 by groups of products

Fig. 13. Imports from outside　　　　Fig. 14. Exports from outside
the Community in 1975.　　　　　　　the Community in 1975.
Total: US$155 mrd.　　　　　　　　　Total: US$155 mrd.

This demonstrates how necessary it is for Europe to maintain its ability to export manufactured products and capital goods so as to pay for the imports its inhabitants need in order to live.

One might take the view that it matters little if this trade is unbalanced; if it is marginal, we should organize ourselves to be self-sufficient. This would be a major error of interpretation. International trade is essential to the economic life of the Community. In 1975 general imports by the Nine represented 22.3% of the gross domestic product (25.5% in 1974), and the share of world trade was high at 37.6% of the total, made up of 18.2% for trade within the Community and 19.4% for trade with the rest of the world. The shares of the United States and the Soviet Union are 12% and 4.6% respectively. If circumstances compelled them to, these continent states could live within their frontiers without suffering too greatly; we Europeans could not.

With whom does this trade take place? The geographical structure of the Community's trade with the rest of the world in 1975 is shown in Table 6.

TABLE 5. Community trade balances with non-member
countries by class of product in 1975
(million EUR)

| CST | | EUR 9 | W. Germ. | France | Italy | NL | Belgium | UK | Ireland | Denmark |
|---|---|---|---|---|---|---|---|---|---|---|
| 0-9 | Total trade | -3 802 | +9 932 | -902 | -2 126 | -3 709 | -1 210 | -4 787 | -382 | -617 |
| 0+1 | Food, beverages, and tobacco | -8 533 | -2 716 | -325 | -1 441 | -1 031 | -537 | -2 752 | -35 | +304 |
| 3 | Fuel products | -31 917 | -6 417 | -7 723 | -6 283 | -3 130 | -1 679 | -5 733 | -196 | -758 |
| 2+4 | Raw materials | -13 588 | -3 861 | -2 127 | -2 400 | -1 034 | -1 018 | -2 894 | -54 | -202 |
| 5 | Chemicals | +7 788 | +3 415 | +1 085 | +711 | +898 | +448 | +1 188 | +33 | +10 |
| 7 | Machinery and transport equipment | +34 469 | +16 777 | +5 525 | +4 294 | +840 | +567 | +6 205 | -95 | +356 |
| 6,8 | Manufactured goods | +8 849 | +3 247 | +2 541 | +3 285 | -244 | +1 065 | -665 | -54 | -328 |
| 9 | Goods not class-ified elsewhere | -870 | -513 | +122 | -292 | -8 | -56 | -136 | +19 | +1 |
| 0-4 | Primary products | -54 038 | -12 994 | -10 175 | -10 124 | -5 195 | -3 234 | -11 379 | -285 | -656 |
| 5-8 | Industrial products | +51 106 | +23 439 | +9 151 | +8 290 | +1 494 | +2 080 | +6 728 | -116 | +38 |

TABLE 6

| Origin Destination | Imports | | Exports | |
|---|---|---|---|---|
| | 1975 | 1974 | 1975 | 1974 |
| Total non-member countries | 100 | 100 | 100 | 100 |
| Of which: Industrial countries (Class 1) | 48 | 46 | 52 | 58 |
| Developing countries (Class 2) | 44 | 47 | 36 | 31 |
| State-trading countries (Class 3) | 7 | 7 | 11 | 10 |
| Other | 1 | – | 1 | 1 |

The relatively high volume of existing trade with the developing countries should
be noted. They are our natural trading partners, as they are partly dependent on
us for their technology, their equipment, and the training of their specialists,
while we depend on them for our raw materials, energy, and tropical agricultural
products. But where is there, under reassuring conditions of stability, a dialogue
on Euro-Arab or Euro-African economic reciprocity? Who can guarantee the permanency
of trade in the long term? Is it conceivable that a European state, however
powerful and determined, can alone control these vital trade patterns when the
developing countries tend to be taken as hostages in the conflicts of interest
between the more powerful?

This economic dependence on the stability of trade patterns, which are beyond our
control, is frightening. It accounts for the fact that, in the eyes of many
Europeans, American protection appears inevitable and is worth much in the way
of concessions.

Although it cannot provide all the solutions, research and development can be a
considerable factor in improving this situation or at the least consolidating some
of the stability factors.

If our industries are competitive and our technologies in the forefront of progress,
they will continue to be in demand by our customers in the countries now becoming
industrialized, where we have the advantage of posing no threat of imperialism.

If technical research is called on to turn our natural resources to the best account
(solar energy, processing of low-grade ores, substitute products), to conserve
scarce products, and to propose consumption without waste, it can help to make
Western Europe less vulnerable.

This analysis compares Europe with Japan. However, the similarity is lessened by
a number of differences: the natural frugality of the Japanese people probably fits
them much better than it does the Europeans for an age of economic restriction.
Japan appears to have found in its deep-rooted culture and traditions cohesive
forces that are lacking in Europe, and has realized unhesitatingly that it must
wage a merciless economic war based on technological progress.

## CAPACITY FOR INNOVATION

As the preceding sections have clearly shown, in a world where mass effects tend to become preponderant, Europe's chances depend mainly on the quality level it can attain.  That is why its capacity for innovation is so important.

For a physicist who reached maturity in 1950 and took a look backward at his past, this question of Europe's contribution to scientific and technical progress would have induced a superiority complex.  Had not the laws of conventional mechanics, relativity, and wave mechanics been discovered in Europe just as the quantum theory had been worked out there and the main experiments in particle physics conducted?  Did not Western Europe supply the lion's share of the inventions necessary for the production and use of electricity, automobile traction, radio and electricity, aeronautics, jet propulsion?  Did not the first nuclear explosion bear the stamp of the efforts and the genius of physicists who came from Europe?  And was not European superiority also decisive in the chemistry of carbon derivatives, in the synthesis of nitrogenous fertilizers, in the production of aluminium and chlorine, in medical asepsis techniques, chemotherapy, vaccination, and serum therapy; were not antibiotics born in Great Britain?  Was not inventive genius in mathematics essentially European?

Twenty-five years have been sufficient to knock down this fine assurance, and although Europe still has little doubt of its inventive capacity, it is already concerned about its capacity for translating the results obtained in its laboratories into industrial and commercial products.[1]

The capacity of the European Community countries for scientific and technical innovation measures up badly today to that of the United States, Japan, and Soviet Russia, with whom Western Europe is competing under various heads, above all if an attempt is made to forecast Europe's future chances.  The time and the means are not available for carrying out a detailed study of this question.  It would be desirable to employ the procedure of a systematic study, a kind of stocktaking, clearly distinguishing the position for each major branch of industry because the situation varies a great deal from one sector to another.  Up to what point should we be concerned at the deficiencies of countries like Federal Germany in knowhow and patent licences or of France in copyrights in the cultural field?

In the absence of a clearly defined study it must suffice to have the results available in the bibliography or derived from interviews with a number of personalities in the scientific and technical world.[2]

If we start at source, i.e. basic research, the general impression is not one of Europe falling behind in knowledge but rather one of lack of concern.  Whereas in the United States, after a difficult period, basic research has got going again, whereas it has never been noticeably braked in the Soviet Union, and whereas Japan has decided to take the offensive, Europe's wealth has been permanently diminishing since the end of the sixties.  There are certain signs of a deterioration: the United States is well ahead in original publications and Nobel prizes although, thanks to the free movement of men and ideas, there is as yet no

---

[1] See Annex II, a list of the main scientific discoveries since the Renaissance compiled by Mr. Macioti.
[2] Survey conducted by FISH (International Foundation of Human Sciences, Paris).

feeling among European scientists of lagging behind intellectually.  One cannot
speak, therefore, of a scientific gap.  It will remain to be seen whether this
comfortable assessment is borne out by facts and especially whether it is not so
that Europe is still living on its past impetus without yet having suffered the
consequences of the reduction in its scientific effort.

If we move on from basic research to applied research, development and industrial
innovation, what we see varies greatly, depending on the speciality.

### *Europe's place in the "power specialities"*

In the "power specialities" such as those associated with political or military
power, armaments, space, aeronautics, and nuclear science, the place occupied by
the Community countries is remarkably satisfactory as regards the qualities shown
by European technicians, although at the same time disturbing because of the
inadequate economic exploitation of successful results.  There are always
obstacles: either the capacity for financing is lacking or the internal market is
too compartmentalized or the outside market is too resistant or the desire to be
in on things on a major scale is not great enough.  The United Kingdom and France
have been able to develop a sophisticated arsenal without outside help: their
electronic industry is not inferior in terms of performance to that of the Americans.
Concorde is the world's leading commercial supersonic aircraft, the satellite
Symphonie is a demonstration of the Community countries' ability to solve the
problems posed by space telecommunications.  Europe is ahead in nuclear breeder
reactor techniques for the production of electricity.  However, despite these
technical successes in the "power specialities", the place occupied by Europe in
the volume of world achievements is remarkably humble.  In commercial aviation,
the place of European aircraft in the world fleet is small (Fig. 15) and Concorde's
experience shows how many difficulties arise in the face of technical superiority
in this sensitive field.

Judged by quantity rather than quality, the inferiority in nuclear armaments is well
known.  The same goes for conventional armaments if we consider the range of
production, despite the fact that for reasons of product quality and also for
reasons of political independence, the European conventional weapons industry is a
great exporter.

In space, Europe's effort is modest, to say the least.  It has played no significant
part in the great prestige ventures of manned satellites, moon landings, and
exploration of the planets by space probes.  Experience has been gained on a
reduced scale, as Table 7 shows.

What guarantee of independence does Europe have as regards major problems involving
satellites such as telecommunications, data transmission, broadcasting and television,
meteorology, remote sensing, vehicle spotting, and military information?  And yet it
has the skills, as is proved by the successful part it has played in some of the
scientific space programmes.

In the production of electricity of nuclear origin, after some countries had
abandoned their original lines, European technicians gave proof of their talent by
regaining their freedom of design and engineering for the conventional types and
taking a firm lead in several major techniques such as fast neutrons, the breeder
reactors.  However, the United States opposition to an independent export policy
is reinforced by demonstrations of concern by environmentalist movements.

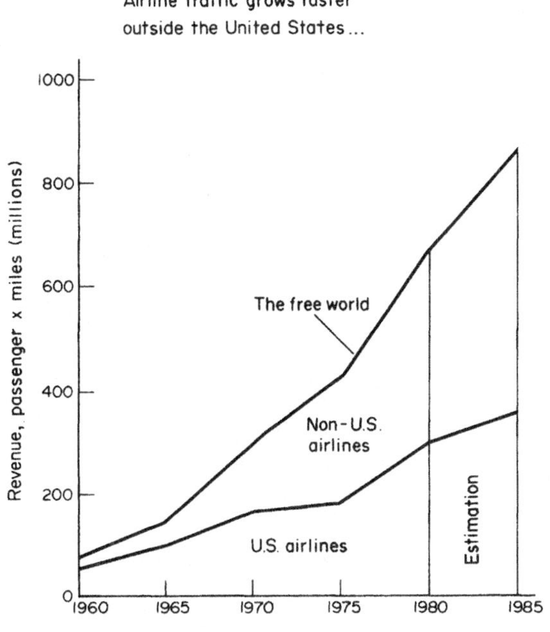


Data: Pratt and Whitney.

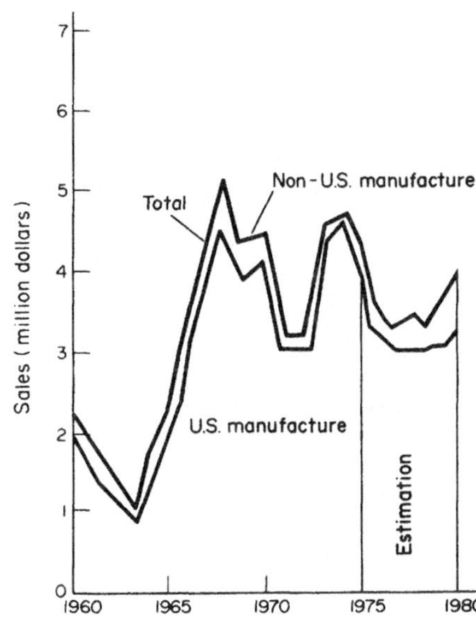


Note: USSR, PR China and non-ICAO
countries.
Data: The Boeing Co.

Fig. 15. Aeronautics (*Business Week*, 12 April 1976).
Growth of traffic and manufacture.

Europe is probably staking its future on the availability of energy, and it is
essential to overcome the existing internal obstacles; they show an inadequate
understanding of the general problems to be overcome and the importance of the
guarantees to be given regarding reactor safety and waste reprocessing - all problems
that are being studied, it is true, both by the Community's Joint Research Centre
and by various national centres. Europe's capacity to shape its own destiny and
accept the risks involved will be revealingly measured by the quality of the effort
it will make to produce electricity of nuclear origin and the nature of the
consensus it will obtain from its peoples in understanding the balance of the risks
accepted, by the level of investment to guarantee safety, and by its resolve to
take part and have a say in the control of nuclear proliferation.

The exploration and opening up of the oceans (collection of deep-sea nodules, deep
drilling, etc.) must also be classed among the power specialities. It is not yet
clear how these will benefit Europe in the future, despite the pioneering successes
obtained. During the twenty-first century, this field could play a similar role,
in the economics of primary resources to that played by the discovery of the new
worlds in the sixteenth century. However, this will call for human and capital
resources obviously beyond the capacity of any individual community country, and
it is to be expected that political obstacles will emerge similar to those
encountered in the past in the control of fossil energy resources or today in
the aviation, space, or nuclear markets.

TABLE 7. Census of objects in space (at 26 December 1976)
Source: North American Air Defense Command of the US Air Force

| Country/ organization | Earth orbits | | Interplanetary space | | Total | Disintegrated objects | | Total |
|---|---|---|---|---|---|---|---|---|
| | Payloads | Debris | Payloads | Debris | | Payloads | Debris | |
| United States | 401 | 2 281 | 27 | 37 | 2 746 | 431 | 943 | 1 374 |
| Soviet Union | 399 | 840 | 25 | 8 | 1 272 | 691 | 3 371 | 4 062 |
| United Kingdom | 7 | 4 | – | – | 11 | 4 | – | 4 |
| Canada | 8 | – | – | – | 8 | – | – | – |
| France | 13 | 42 | – | – | 55 | 1 | 23 | 24 |
| ESRO | 1 | – | – | – | 1 | 6 | 3 | 9 |
| Federal Germany | 2 | 4 | 2 | 1 | 9 | 2 | 1 | 3 |
| Australia | 1 | – | – | – | 1 | 1 | – | 1 |
| Japan | 8 | 8 | – | – | 16 | – | – | – |
| P R China | 4 | 4 | – | – | 8 | 3 | 9 | 12 |
| NATO | 3 | – | – | – | 3 | – | – | – |
| Netherlands | 1 | – | – | – | 1 | – | 3 | 3 |
| Spain | 1 | – | – | – | 1 | – | – | – |
| France/Germany | 2 | – | – | – | 2 | – | – | – |
| India | 1 | – | – | – | 1 | – | – | – |
| Italy | – | – | – | – | – | 4 | – | 4 |
| ESA | 1 | – | – | – | 1 | – | – | – |
| Indonesia | 1 | – | – | – | 1 | – | – | – |
| Total | 854 | 3 183 | 54 | 46 | 4 137 | 1 143 | 4 353 | 5 496 |

## *Europe's place in information technology*

We shall see later how important everything connected with information is for the
activities of the post-industrial society.  It is therefore essential for Europe to
be capable both of creating new technologies for the recording, transmission, or
processing of information and – even more important – of making good use of them to
improve the productivity of its services and industries.  These new tools have to do
with the human mind, intellectual processes, and relations between men, and it is
therefore necessary to make their use consistent with deep-rooted cultural forces.

In this very complex field, Europe has not got off to a good start by comparison
with the United States, and is tending to be overtaken by Japan.  For the first time
in recent history, with semi-conductors, integrated circuits, and computer software,
it is in the presence of technology which does not owe its invention mainly to
Europe.  We must be careful not to exaggerate this verdict, derived from statistics,
because it does not take into account certain possibilities of a reversal of technical
and trade positions.  Technical innovations in this particularly unstable field can,
of course, provide surprises where the Community is still capable of seizing the
initiative.  However, the facts are impressive today, as is shown by the graphs
which follow.

As regards the world breakdown for the installed value of computers for 1974, which
constitutes a yardstick for computerization, Western Europe occupies second place,
far behind the United States (Fig. 16).

If we now examine trade penetration for the supply of calculators in national
markets we see a major difference between the European situation — extremely open to
imports or to the multinational companies of US origin — and the Japanese situation,

where there is a tendency to reserve for Japanese nationals the lion's share of the
market[1] (Figs. 17 and 18).

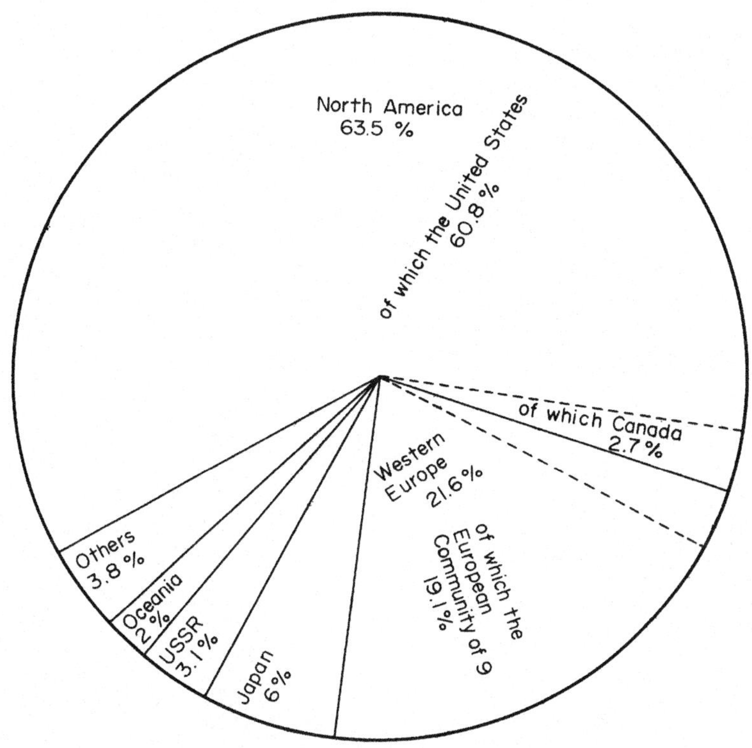


Fig. 16. World breakdown of installed computers in
terms of value by major geographical areas.
Source: Japanese Computer Industries –
SRI – USA, 1974.

A study should be carried out to ascertain the situation and trends as regards
microprocessors, as these small devices may be regarded as the nerve cells of
tomorrow's society.  Distributed very widely, they will be used in all arithmetic
operations required for trade, industry, and accounting purposes and also in all
intelligent automatic systems in which they will add a memory and logic capability
to the mechanical automaton.  Combined with quartz they supply solutions like
clockwork, combining extreme precision with low prices.  These microprocessors are
the result of complex physical processing operations, which means that it is
economic to produce them in large numbers; female Asian labour is particularly
suitable for the delicate assembly operations.  In the international distribution
of labour, which is unfolding before our eyes.  Europe's chances would appear to be
poor in comparison with US and Japanese companies, which subcontract certain
finishing operations in the various South-East Asian countries open to a free market
economy.

---

[1] These graphs do not take into account changes which have occurred in France since
the CII Honeywell-Bull merger.

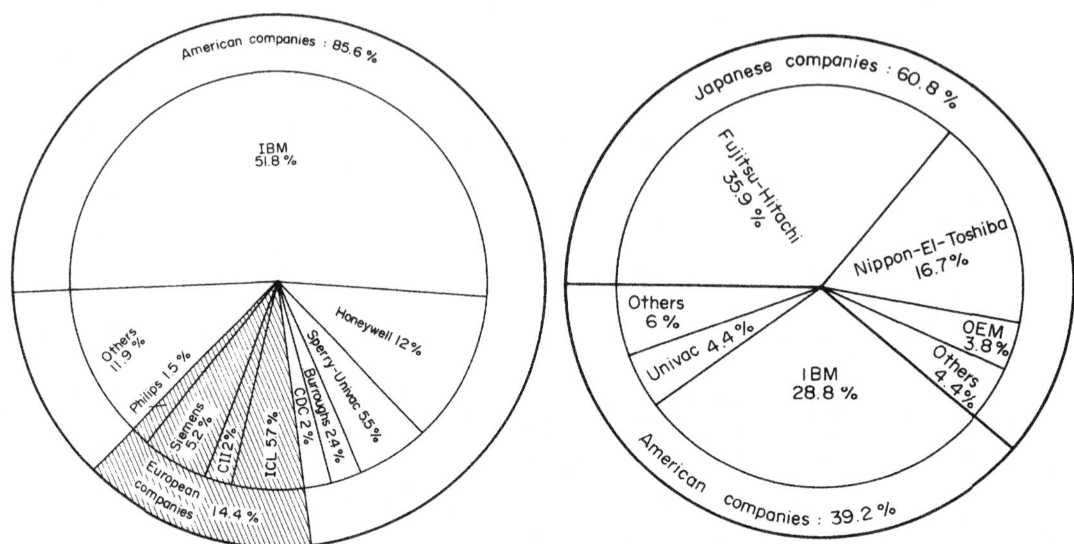


Fig. 17. American computer companies on     Fig. 18. American computer companies on
the European market in 1975                 the Japanese market (1974).
(in terms of the value of                   Total: $44 billion. (Source:
equipment installed). 1975:                 MITI.)
Total: $19 billion. (Source PAC.)

The difficulty experienced by the European Computer industry in expanding may be
put down to the excessive power of a multinational company of American origin that
has acquired a *de facto* monopoly position as regards medium-sized and large
computers. This explanation does not apply in the small computer sector, where
handicaps resulting from scale effects are less serious. The information in
Fig. 19 (even allowing that another source gives a slightly different picture)
shows how small is the share acquired by European companies in a field vital for
the automation of production and the introduction of data processing into all
disciplines.

*Europe's place in the traditional industries*

Europe's capacity for transforming innovation into commercial success encounters
fewer obstacles in the traditional industries, especially in those where Europeans
have been influenced by their own characteristics. One example is the motor
industry, well geared to the conditions prevailing on the home market. The
dimensions of European cars, their comfort, and their fuel consumption are suited
to the small confines of Europe's territory and the need to save energy. As these
conditions are also encountered in many outside countries, the Community's car

industry is a great exporter. The taste for quality is satisfied by colour television standards which were defined later than American standards but are superior. The development of health schemes has been beneficial to the pharmaceutical industry and certain specialist branches of production such as deaf aids. The German and Swiss chemical industries are predominant in some techniques, and The Netherlands has an outstanding reputation in electronic equipment for civil use. Further examples could be added in this way to demonstrate that there is not really a technological gap.

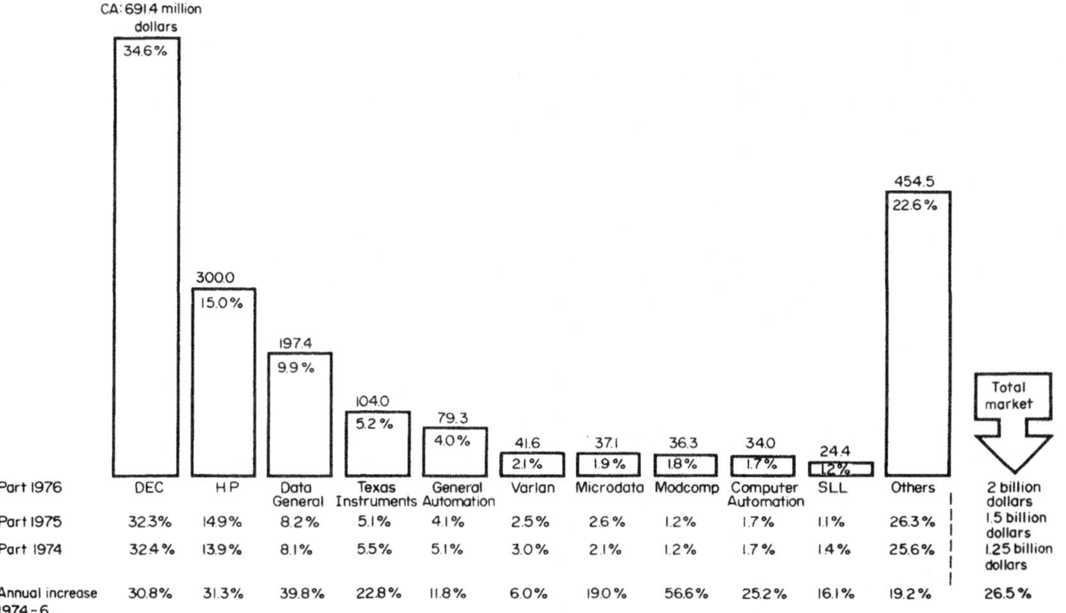


Fig. 19. Estimate of the shares of the world market for minicomputers accounted for by the main manufacturers. Source: Robertson, Colman, Siebel, and Weisall.

However, the Community countries will in future have to keep an eye, even more than in the past, on the potential for competitive innovation of Japan, which, in the traditional industries, is putting up a redoubtable performance, especially as regards the automation of production (steel, shipbuilding) and very large-scale mass production as in the case of optical instruments, photographic and film equipment, radios, tape recorders, portable television sets, reprographic equipment, motorcycles, and cars.

This need to be more inventive as regards the automation of mass production also appears to be a major point in the textile industry, in the processing of plastics, and in certain agricultural and foodstuffs specialities in countries where the fruits of low-cost labour using conventional means may be the ruin of our industry. If domestic production were to disappear almost entirely, it is not certain that the capacity for designing, manufacturing, and exporting production equipment would survive.

If considerations of political independence could be disregarded, the basic industries, e.g., cement works, paper mills, sugar refineries, oil refineries, chemical factories for the production of plastics, manmade textiles and fertilisers,

and aluminium works, would be better sited in areas where energy is cheap or where
ore is rich or where the low population density reduces pollution hazards.  Their
continuation in Europe cannot, therefore, be taken for granted in the context of
an international division of labour resulting from the free movement of products
manufactured on the most economic sites.  Only innovation making it possible to
reduce the effect of pollution, to exploit low-grade ore, or to economize energy
will enable them to continue on a competitive basis.  This battle is far from won
and a social study would be necessary in order to find out to what extent it is
being waged.

To sum up, Europe is very badly placed as regards innovations in the power
speciality fields, is behindhand in information techniques, and is also threatened
as regards its capacity for innovation in the field of traditional industry.
However, the talent and the knowledge of its specialists constitute forces which
are apparently intact.  Up to now the scientific and technical objectives assigned
have been met.

Upon a number of occasions in the course of this study we shall be coming back to
this question mark concerning Europe's capacity for innovation.  We shall see, as
all specialists in development, demonstration, and industrialization know quite
well, that the principal deficiencies are by no means a lag in knowledge.  It
might be thought that this leeway could be made up by the purchase of licences or
transfer of technology.  On the contrary, these facilities contain the risk of
deepening the gulf between the donor and the assisted party since the assisted
party makes a saving and is ignorant, therefore, of the main part of the inventive
process, consisting of the effort involved in adjusting the new product to the new
market and the always gradual discovery of technical solutions which govern the
lowering of the cost price.  The assisted party loses the initiative as regards
laying down specifications for the properties of products and their use.  These
are decisive factors for the acculturation of new technologies.  The assisted
party has, as it were, his inventive genius amputated and is unable to wage a
battle on a moving front.  It imports technique which is not perfectly adjusted
to the peculiarities of its market.  In this way, without any immediately perceptible
warning, it slips into underdevelopment.

If, therefore, there is no alternative to the importation of technology, in order
to make up leeway, it must be realized that, to avoid a situation of permanent
dependence, the research and development effort devoted to the field concerned
must be at least equal to that of the licensee whose knowhow is being purchased,
so as to counterbalance its influence in the initial stages and later to become
independent of it.  This method of regaining one's scientific and cultural identity
in respect of an imported technology has been Japan's strength, and there are
numerous European examples, especially in Germany.  The temptation to adopt the
opposite attitude is great, however, and this is where the danger lies: it is all
too easy to fall into the comfortable (for the time being) situation of the
hanger-on.  Those who have not had personal experience in handling these matters
are ill-equipped to understand the importance of access to knowledge and the far
greater importance of the capacity to acquire knowledge by one's own efforts or
to add original elements to it.


## THE CHALLENGING OF MORAL VALUES

There is a crisis of moral values in all industrialized countries.  Europe is no
exception, and it is difficult to know whether these phenomena are really deep-
rooted or whether the descriptions of them are no more than the expression of a
mental susceptibility on the part of a fairly insignificant section of the adult
and adolescent population spoilt by the comforts of the civilization of plenty.

It appears as if a period of intoxication in which the favourable aspects of
technical progress were prized to excess has given way to an exaggerated movement
in the opposite direction which asserts that all is harmful in the consumer society.

Whatever the underlying reasons this crisis must be taken seriously, however, as it
can sap our moral forces, those forces based on common spiritual values which nurture
solidarity, hope and enterprise and forge the social consensus.

* * *

If this description were taken to its limit, verging on caricature, the following
picture of this crisis would emerge:

(1) Youth no longer appears to believe in its future.  Young couples reject life
    and no longer ensure that their generation is replaced.  The need for a
    sustained effort during the period of higher education is grudgingly accepted,
    and badly motivated students find selection by competitive examination not to
    their taste.  Sexuality is no longer, as it was in the course of preceding
    centuries, an occasion for asceticism or faithfulness to the indivisibility of
    the couple.  On the contrary, it is the occasion for manifesting a capacity
    for liberty.  More and more people are turning to alcohol and drugs.

(2) Certain values which counted as fundamental for previous generations have been
    called into question.  The family is losing part of its stability.  Religion
    is rejected, especially as regards its aspects of moral constraint and
    institutional organizations.  There is controversy in the churches, and the idea
    of patriotism has ceased to be felt my many, while the spirit of national
    solidarity is yielding ground to regional movements.

The image conjured up by the pessimistic observers of industrialized civilization
continues as follows:

(3) The man on the job is not happy.  Senior staff feel they are lost in gigantic
    organizations which have become too complex for them really to participate in
    decisions.  A country's own people reject manual tasks, which they entrust to
    immigrant workers cut off from their family and cultural environment.  Workers
    have the impression that they are no longer recognized as people but are
    regarded rather as things; the dignity of labour has been lost.  Power sharing
    is not fruitfully determined between management, the unions, and government.
    Ideological opposition is added to the natural difficulties involved in the
    period of apprenticeship in the very complex relations imposed by the
    organization of major enterprises.  The fear of unemployment sets in.

(4) The townsman is not happy.  The town is ugly, inhuman, and dangerous.  Far too
    much of his free time is taken up by travelling.  Contact with nature is lost.
    Communications between human beings are barren and anonymous.  The lack of
    space in the home makes it difficult to exercise an art, to practise a craft,
    or just to be on one's own to think.

(5) Man is no longer "sapiens" or "faber"; he has become "consumens" and is no
    longer able to resist the temptations of consumption.  He becomes indebted in
    order to buy useless objects which do not contribute to his happiness.  He is
    no longer able to "be", so much does he wish to "have".  His pleasures are
    passive.  The greater part of his leisure time is taken up by the television
    set or the motor-car stuck fast in traffic jams.  The radio and the press
    spread, above all, bad news.

(6) The citizen prefers security to a spirit of enterprise.

(7) The State tends to be regarded as all-providing.  It is responsible for
everything which is difficult or uncertain.  It guarantees order and national
defence and manages communications.  It must supply family allowances, health
and old age insurance, guarantee permanent employment and pay the unemployed,
subsidize agriculture, dispense health care and education free of charge, and
control industry make it competitive and where necessary make up losses.  It
is responsible, in both the short term and the long term, for inflation and
the external debt, for scientific and technical research, and for cultural,
artistic, or sporting events.  However, those who demand these measures find
the taxes too heavy and individual liberty pared down too much.

(8) Violence is increasing.  Juvenile delinquency, kidnapping, ransom demands,
skyjacking, and bank robberies are becoming more and more frequent.

It would be unwise to exaggerate the importance of these symptoms, but it would
also be rash to take them too lightly.  It cannot be denied that these difficulties
are being experienced to some degree by society and that the very discussion of them
will aggravate the situation if the necessary answers are not found.

This malaise, the real depth of which is not at all clear, is the outer crust of
the psycho-sociological crisis which Europeans are going through in adjusting
themselves to a world which is changing more and more quickly and is out of their
control under the pressure of technical progress.  This bewilders them, not because
they do not understand the intellectual machinery involved but because it destroys
their balance and their habits.  Well established yesterday on his land and amid
his traditions, the European feels himself today uprooted in towns which he no
longer understands.  He does not know what social attitudes constitute the right
response to these material changes.  He is discovering that he is no longer the
driving force and the author of history, and it goes against the grain to be
nothing but the vassal of "imported" history where others decide and where, at best,
he can only run along behind.

In the enthusiasm of the post-war reconstruction effort, followed by the almost
fetish belief in the virtues of industrial growth, Western Europe has completely
transformed itself.  However, this rapid, brutal, exogenous trend in history is,
for the European, a challenge for which his culture has hardly prepared him and
he cannot see himself in this materialist role gorged with manufactured products,
that it would impose on him.  He is starting to denounce the art of economics; all
this growth seems to him to be out of proportion with the human scale which his
landscapes and his experience have given him.  Memories and scruples are mingled
within him and conjure up a nostalgia for yesterday's society.  He recognizes it
to have been elitist and regrets this but nevertheless attributes to it much
variety, many virtues, and human values.  In the mass society which he desires and
yet fears he discerns not a new Athens which would be justified and which would
enchant him but a new Sparta with threatening conformity, with neither privilege
not individuality.

He remembers, above all, that the Renaissance started off the rational transformation
of the world, which is being completed before his eyes, by launching Europe on the
road of the systematic and continuous progress of knowledge which for some centuries
was to guarantee its pre-eminence.  The Renaissance did this, more particularly, by
instituting "bachelors", that surplus of students which the universities trained and
awarded diplomas without guaranteeing them, within themselves, either a place or a
career.  This radical and emancipatory initiative launched the new elite, at their
own risk and peril, along the path of scientific, technical and even social innovatio
A new spirit — the spirit of enterprise — was thus introduced into the advancement of
knowledge and was to fertilize it.  It is to this spirit that we owe the transition
from conformity in accordance with a single thesis which we see incarnate in the

word "university" to a pluralistic situation.  Europe is well aware that it owes
its glory to this, and that this is still its genius.

In entering upon the twenty-first century it is to be hoped that Europe will
experience a "Second Renaissance", but this will not be possible without intellectual
and social boldness.  In particular, is not an important stage the assimilation of
innovation by our culture, i.e. the integration of change into our society?  It
would be necessary, while abandoning all the old quarrels about elitism and access
to knowledge, effectively to bring to life again our capacity for assimilating
novelty.  If it is preferred, this may be expressed as our "pedagogical function"
of implementing knowledge, more particularly knowledge in the field of the human
and social sciences, still so desperately trivial by comparison with the "exact"
sciences.

In this crisis of moral values the European Community is now apparently characterized
by mixed reactions.  Some countries, economically and socially, seem to be suffering
more than others.  Having entered the crisis earlier, are they not that much nearer
recovery?  It would be wise to recognize that, overall, there is similarity in the
dangers and that the discrepancies in behaviour do not indicate balances which have
stabilized differently but, more probably, that certain thresholds have been crossed
which everyone is near to — only some are still on the right side and others already
on the wrong side.

The above description should not be allowed to discourage us.  In this last half of
the twentieth century there have been no extreme innovations as regards immorality,
abuse of power, or alcoholism.  Historians teach us that, even though violence has
increased since the abnormally low level observed in the early decades of the century,
it is still nothing like the violence that prevailed throughout most of our history.
Illiteracy has largely disappeared together with famine and epidemics; extreme
poverty is far less common, and instead of fighting each other in fratricidal wars
the European nations are living in peace and seeking co-operation.  The debit balance
in terms of the rebellion against systems of moral values must be set against the
almost triumphant achievements in public health, in the increase of purchasing power,
in the eradication of poverty, and in education.

Cannot the crisis of moral values in the developed countries be interpreted as the
reverse of the coin, the unfavourable aspect of a melting-down phase in preparation
for a further adaptation?  The right side would then be the readiness, as a result
of this break with the past, to accept the construction of a new future for which
only one project is still missing — the Second Renaissance.


## WHAT ARE THE CONCLUSIONS OF THE "BALANCE-SHEET"?

Within the limitations of probability for the figures advanced, it is possible to
draw from the above balance-sheet a number of principal points:

(1) In the space of a generation, Europe's place in the world has changed radically.
    It no longer controls more than one-third of the land and one-quarter of the
    peoples together with their wealth in the form of primary resources.  The
    future will be determined by this change which is still masked, in the current
    transitional period, by certain trading habits and cultural traditions.  In
    particular, Western Europe no longer has the capacity for taking the initiative
    based on political power.  It is in danger of being overtaken by events instead
    of being their source.

(2) In terms of its relative importance in the world population, Europe is in a
continual state of decline.  Towards the year 2000 the Community of the Nine
will be smaller, in proportion, than one of its major members in the eighteenth
century.  The ageing of the population will then contrast with the youth of the
southern countries where numbers will have become greater.

(3) Through its geography and climate, Europe is to be distinguished from the
continental states such as the United States, the Soviet Union, and China, with
whom it is competing.  Restricted as regards the amount of space available,
short of rich ores and fossil energy, and fragile as regards its ecology,
Western Europe is ill-suited to giant-scale industry and cities.  It may be
distinguished, however, from Japan by its cultural traditions and the wide
range of factors which make for diversity.

(4) Europe is threatened with permanent scarcities which it will be able to meet
only by means of imports, which renders it economically and politically
dependent.  The field where it is most sensitive in this respect is energy,
where the situation may become dramatic.

(5) Compelled as it is to export in order to pay for its purchases of basic
products, Europe is asking itself what is left to it of its capacity for
innovation, its scientific and technical primacy having first been challenged
fifty years ago.  It has lost the initiative in armaments and in space and is
in danger of losing it in the exploitation of the oceans.  Its handicap in the
field of information technology is worrying, and when it does achieve breakthroughs in aeronautics or the nuclear field it has difficulty in exploiting
them commercially.  However, the competence of its scientists and engineers
seems to be intact and leads to success whenever problems specific to the
European market are dealt with.

(6) The system of values which constituted Europe's moral strength is being
contested.  This crisis is the manifestation of a failure to adapt culturally
to the changes caused in society by the brutal advent of the new techniques.
This crisis, which is affecting the whole of the industrialized world, is
particularly felt in Europe.  But it can be regarded as a preparation in
readiness for a project, a Second Renaissance.

Chapter 2

# THE CULTURAL FORCES

The major changes in the situation which have taken place since the Second World
War must be seen as facts.  It would be futile to regret them, or even to look for
the source of the human responsibilities.  The obstacle to be overcome is rather
that of accepting them as being true.  We are still attached to sentiments derived
from a previous stage in which Western Europe was made up of an assembly of powers
which were great because of their military and political strength and were wealthy
in terms of human beings available for the remainder of the world.  The peoples who
recently became independent continue to look at us partly through the eyes of the
past, and the United States, instead of fearing the complications which might ensue
from our weakness, still consider us as trade adversaries.

Part of the influence which Europe still has today is linked with the time when we
were still formidable to others.  Part of the freedom with which we give vent to
the expression of our national selfish feelings is derived from the fact that we
have not become aware of our true standing in today's world nor of the dimensions
of the problems to be solved.  Our minds may be informed of the new facts while
our sentiments may act as if only the facts of former times counted.  When, between
these two aspects a mutation takes place, misinterpretations are easy, and this is
what must be avoided.

This part of the study constitutes an attempt to define the ideas and forces which
emanate from European culture and the aspirations which may be deduced therefrom.
Europe must not be static, but growth can no longer be anything more than
essentially qualitative.  In addition to the factors of evolution inherent in the
various economic situations there are underlying psychological forces which, there
can be no doubt, must now be clarified and used as a guide in rechannelling
technological evolution.

* * *

The overriding idea in this chapter is that Europe's almost sole wealth lies in the
quality of its inhabitants who are enclosed in a culture.  This culture is part of
history, and it is also part of the climate, the scale of nature, and the absence of
easily accessible great material wealth.  Because the confines of Europe are narrow,
the density of population is great, and the level of education is high, contact
between minds — and clashes between them — is facilitated.  It is these factors of
density of population, ease of communication and high level of education that help
to maintain the quality of man in Europe.

This culture is acceptable in the outside world because the tide of expansion of
recent centuries has left behind it languages and systems of education of
European origin even where the peoples are not the children of immigrants.  This
culture is still regarded as powerful and useful by the recently independent
peoples who, in order to assimilate the new technologies, need to take as their
inspiration the classic European models in the same way as the Middle Ages and
the Renaissance were helped by the Greek and Roman heritage.[1]  It is, moreover,
this very profound cultural interpenetration with certain regions in the world,
especially North America, which makes it so difficult to recognize that Western
Europe has its own cultural identity.

The same problems of assimilation of scientific and technological progress appear
in the various industrialized countries, and the same solutions will in the end
be applied, but Europe and Japan are the only countries which are today impelled
by necessity.  The major gain of the last fifty years was the growth in capacity
for production and consumption of material goods on the part of agriculture and
industry.  This growth consumes a great deal of space and primary resources and
gives rise to nuisances.  While the available space, energy, and raw materials
were considerable and while the nuisances could be absorbed by natural recycling,
there was nothing to justify such growth being identified with waste.  It was, on
the contrary, legitimate to consider it as progress.  It makes available to man
wealth which is useful from the point of view of his material and immaterial needs.
It is the sign of man's victory over nature.  The United States, Canada, and Russia
(with its Siberian appendages), Australia, and Brazil have no reason in the short
term to conduct a vigorous quest for another form of balance.  Their natural bounty
grants them safety margins which are still considerable.

Europe and Japan are no longer at this stage.  They are nearing the limits to the
exploitation of arable land, to population density, to the exhaustion of their
soil in terms of wealth in primary products, and to the elimination of nuisances.
In this they are close to the most densely populated developing countries.  What,
in the new countries, is a conquest over nature is for them waste, and they have
lost any chance of imposing politically their complementarity upon less-exhausted
regions.

As always, when a civilization shows all the signs of success as regards prestige
and power, the American model fascinates.  Performance is measured by the amount of
the gross domestic product, for a whole country or *per capita*, and the aim of all
is to come as close as possible to the American figure.  And yet the GDP gives no
indication of the quality of life and has virtually no meaning, even in econometric
terms, in the developing countries where part of the wealth is produced by family
or craft activities and cannot be precisely recorded.

It is physically impossible, at least for the next thirty years, to extend average
American consumption to the world as a whole.  Capital formation is not fast enough
and the available resources are not adequate.[2]  In view of the short time available
to find solutions to the problems that will be caused by a population of 8 to
12 000 million next century, it appears wrong to pattern the future on the existing
consumption model of the industrialized nations.

---

[1] This idea was expressed by Mr. J. Stoezel, Chairman of the Human Sciences
Committee of the Commission Nationale Française pour l'UNESCO.
[2] For energy, forecasts for the year 2000, taking the extreme hypotheses, give a
level of consumption between 78 000 and 130 000 m. tpe, about twenty times the 1972
consumption (5 067 m. tpe), if the American average were extended to the rest of the
world.  (Source: ECA document, Paris, June 1976 "Energy, Growth and Atomic Energy".
m. tpe = million tonnes petroleum equivalent.

On account of its high level of development and relative poverty in terms of space
and resources, Europe is the first of the "rich" regions to encounter these finite
boundaries.  It is obliged to innovate as regards civilization, in order to satisfy
the desire of a population which is not frugal, by proposing new aims for society
that are compatible with its own level of resources and its degree of economic
dependence.

In order to attain these aims it must show originality by comparison with the US
model, which is dragging it in the direction of unsuitable solutions.  If Europe
were to become a sort of guinea-pig region for the world, its absence of an
imperialistic capability and the help it could offer in the form of its cultural
bridgeheads would enable it to convince the financially rich but not yet developed
countries to adopt this new model instead of the model based on growth in the
consumption of material goods.  The United States vitality and potential for
adjustment are such that it could take over and develop on a wider scale original
successes scored in the European social and economic experimental laboratory; in
that case the whole world would get along better.

The modern world is characterized by the interdependence of and consequently
communication between economic systems and the contagious effect of ideas.  It would
be entirely wrong to believe that Europe could isolate itself in order to escape
what it might consider a mistaken trend in other countries, especially the United
States.  This is why nothing is proposed here that is incompatible with certain
trends observed in industrialized nations, especially across the Atlantic.  However,
consideration must also be given to the pressure that will be exerted by the
developing countries in the face of their difficulties; any project that did not
help to reduce this pressure would be neither useful nor realistic.


## THE IDEA-FORCES OF WESTERN EUROPE'S CULTURE[1]

There are no criteria for classifying cultures.  In endeavouring to explain the
principal features of Western Europe's culture we do not by any means claim that
it is better than any other.  Depending on climate, population, and traditions,
other cultures have been born which provide a better response to local conditions
and perhaps even to those for mankind in general.  It would be desirable that, in
the world's major homogeneous regions, such cultures should inspire objectives for
development and serve to consolidate scientific and technical achievements and the
new technological tools.

It appears that, with the possible exception of China, no region in the world has
really tried to control the nature of its economic growth and its technological
aims by making use of the wealth of its cultural identity.  And yet there are
elsewhere considerable philosophical values and intellectual riches, especially in
regions where the peoples and their traditions are most ancient: in the Arab world,
Iran, India, or Indonesia, to say nothing of the original features of the Negro
culture or the remaining influences of the ancient Amerindian civilizations.  None
of these values is being called on, however, for the acculturation of technical
progress.  Japan is a special case: it has managed to industrialized by developing
the same products as the most advanced nations but without abandoning the main lines
of its ancestral way of life and thought.  This is most probably to a large extent
the secret of its success.

---

[1] This chapter, like the previous one, is based mainly on a study by FISH (Henri
Cavanna) and discussions at the Luxembourg symposium organized in October 1976 at
the request of the Commission of the European Communities (DG XII).

Since it is a question here of imagining a new model for society in Europe, it is
as well to attempt the exercise of trying to bring out the ideas-forces which
constitute the basis of Western Europe's culture.  The problem is that of high-
lighting what unites Europeans, the fundamental cultural values they have in
common amid all that divides them in the shape of the variety of ways in which
they express themselves and act.  We shall revert to this later.

Western Europe is a meeting point of cultures.  The Greek pattern of art and
philosophy, the Hebrew heritage of religion and the Jewish diaspora, the Latin
influence, particularly through Roman law, the continuous exchanges with the Arab
world in the course of confrontation or co-operation, have all intermingled with
Celtic and Nordic traditions to produce a "homo occidentalis" who finds it difficult
to identify himself, aware as he is of these differences, even though he contains a
set of likenesses that may be said to constitute his invariants.

What best characterizes Western man is probably readiness for action, participation
in the turn taken by events, and the rejection of resignation.  The centre for new
ideas and growth of techniques was situated first of all in the Middle East and
Greece, and we are the heirs of this Greek-cum-Semitic civilization.  Then the
initiative moved westwards, shifting at the time of the Reformation from Europe's
Mediterranean axis to its north-west axis.[1]  The European, as if impelled by some
force from within, tries to act and to develop through action.  Hence the concepts
of the value of work and the satisfaction derived from abundance to which the
European is today very much attached.

For a long time religion played a vital role in the forming of minds, although its
penetration was not equally intense in the north and south of Europe.  It is
therefore necessary to understand the cultural message inherent in Christianity.
This is not a static religion.  The Holy Trinity is a god in a state of tension,
essentially dynamic.  That God made man means that man is God's partner in creation
and is responsible, alongside Him, for evolution.  Man is not religious only when
in contemplation.  He is religious also in what he does within a framework of
personal freedom which grants him wide scope for interpretation.  Man is great via
his freedom.  Man has a value not only as a species.  Each individual is the
recipient of God's infinite love.  Man is not embedded in castes — his place in
society is only secondary.  The poor man pure in spirit is the lamb of God.  Man
is worth only what he gives to others in God's light.[2]

By projecting these values into the economic sphere, the Reformation fostered the
spirit of enterprise and the reinvestment of wealth.  This movement spread throughout
the north and north-west of Europe and was particularly successful on the North
American continent.  On the other hand, the cultural content of the Counter
Reformation, which was to actuate the Catholic Church up to Vatican II (1962),
partially accounts for the industrial and commercial backwardness of the Latin
countries, the other factor being the shortage of energy (coal).

Dechristianization, which started in the Enlightenment, and the shorter period of
religious impregnation in the Scandinavian nations, did not prevent the establishment
of a certain idea of man inherited from the past and common to all Europeans.
This common idea strongly retains the concepts of dignity, liberty, and
responsibility attached not only to human groups but also to each individual.

---

[1] See works of Pierre Chanu.

[2] Essentially the same values are found in Judaism and Islam, a civilization based
on the Bible like Christianity.

The value attributed to work is profoundly rooted in the popular conscience.  The sense of effort is inculcated from childhood onwards.  The concept of liberty has not ceased to progress in the midst of chaos, becoming an essential value and one which justifies the gift of life.  Western Europe, which has tried everything in its long political history as regards methods of government, fundamentally rejects despotism and demands protection of the citizen by the law.  Such liberty makes possible the expression of Europe's variety of languages and customs, which characterize the mosaic of regions and nations.  Hence the importance attached to the Charter of Human Rights, the free movement of men and ideas, respect for religious or political opinions, and freedom of speech.  To protect his freedom of thought, the Western European invented the separation of Church and State, contrary to developments where the Byzantine influence has continued.

However, the European is also a social creature.  He knows that his success can be the result only of acceptance of a certain number of rules which permit and guarantee life in common.  He wants justice that is not arbitrary, that is based on a law common to and respected by all.  He wants to be governed democratically within rules that are sometimes implicit but generally explicit, within the framework of a constitution.  He demands the right to work and the abolition of excessive inequalities.

The European is firmly anchored in a permanent and subtle interplay between freedom and reason — freedom which would grant him the right to do anything, even what is asocial, or even deliberately to be a cheat or libertine.  Reason, which induces him to give a privileged place to aspirations towards an organized life, makes him respect the civic spirit, gives him a taste for effort, and, at certain times and in some countries, encourages asceticism and puritanism.  This conflict between liberty and reason takes different forms in the north and south, but is ever-present, and its brutal or excessive arbitration is never tolerated for long.  In short the European is an idealist, very much incarnate, who accepts this as one of his fundamental values.[1]

The incarnation is a biological one.  It comes from natural selection, aggressiveness in the struggle for life and evolution, with the elimination of the weakest by the strongest.  This interplay has for a long time made of the European a warrior — which he may no longer be — and the cry for liberty is often confused with this desire to be able to give vent to aggressive strength, either individually or in groups.  However the idealism compensates, in a permanent state of internal tension, for giving vent to aggressive feelings.  It imposes a minimum threshold of justice and compliance with a set of social constraints.

How are the subtle interplay between freedom and reason, this need for individualism counteracted by the concept of the general interest, this need for aggressive action, tempered by the social conscience, understood in our society, transformed as it has been by technological progress?  How can these tensions between humanism and the spirit of enterprise find expression within the complexity of our post-industrial social setups?  All these are fundamental questions to which it is perhaps premature to try to find a reply.  However, the result of this analysis is that no blueprint for society can be proposed to tomorrow's Europeans if it does not give some scope to the biological forces of growth and the spiritual forces of the civic spirit, and if a wide margin of freedom is not granted to the permanent and painful quest for a state of balance between social idealism and the right to be aggressive.

--------

[1] The historical behaviour of Europeans, cruel in their internal wars, destroyers of remote civilizations and sometimes guilty of deplorable mass crimes, does not invalidate the aspirations to idealism, to moral and aesthetic values.  It does, however, reveal the enormity of the contradictions in human nature.

This represents a pull in the direction of progress or, if a less-optimistic expression is preferred, a pull in the direction of adjustment to environmental conditions.

Like all the civilizations that preceded our own, Europe has until now been an elitist society. The European elite, which gradually expanded to include a part of the upper middle class and then more recently some of the middle class, has acquired a highly developed sense of material pleasures as a result of the abundance of agricultural and industrial goods. It had also learned to live amid cultural wealth, in an intellectually and artistically sophisticated Universe, among old stones embellished with ornaments and rich in history, and in luxury of the spirit. The sensuous values impregnated life. The European's contact with the treasures of painting, sculpture, or music; the fact that he moved in cities, which were themselves museums, and prayed in churches, which were art treasures; and the potential of literary production available to him or being created, developed the aesthetic feelings of a growing proportion of the population. Europe spread forth its art and culture. It maintained its diversity. Must not any blueprint for Europe incorporate, as a useful force, these spiritual needs while extending enjoyment of them to the greatest number without, as in the past, reserving them for a minority?

Without expressing this aspiration in formal fashion, the European is manifestly looking for a form of society giving to all what was formerly reserved for a few. This underlying desire in the direction of egalitarianism from the top downwards has many manifestations. It is at the root of the claims of the workers' unions. It explains the desire on the part of personnel in undertakings to be associated with the workings of management. It partly justifies the desire for over-consumption. It is apparent in the form of the refusal to carry out manual tasks. Europe is dreaming of a society in which the class struggle will have become devoid of content because the disappearance of elitism will have caused the solidarity of the privileged few to extend to all the citizens. However, this aspiration is blocked by social structures, certain excesses by way of social inequalities, and general incomprehension in the face of economic and technical constraints. The European does not understand the society in which he lives. He does not tolerate its limitations, its complexity, and its interdependence. He is afraid of social innovation. This gives rise to conflicts exacerbated by support for opposing ideologies, stoked by misunderstandings resulting from language differences, fanned by the conflicting experiences of the various social groups, each of which has frequently lost any sense of the general interest.

These conflicts may be regarded as a symptom of decadence, a weakening of cohesive forces, but they may also be interpreted as a painful and difficult effort to move in the direction of social progress, towards which opponents are tending to converge without realizing it themselves. Is the objective of a genuinely non-elitist civilization, less excessively unequal than all its predecessors, where the most humble would feel that they were recognized and given consideration, a utopian ideal, a dead end, or the great hope of the human race?

Depending on how pessimistic or optimistic he may be, the observer could choose any one of these possibilities, but Europe appears to be making more headway in this effort than the United States, perhaps because its climate, dimensions, and traditions are less virile, less brutal, its peoples are more concerned, and the feeling that an era of economic difficulty is approaching is more acute and more alarming.

## ATTACHMENT TO DIVERSITY

Diversity is a feature of Western Europe.  The ancient Celtic, Scandinavian, Germanic, or Latin cultures have persisted to a degree that varies from one place to another; the influences of the Greco-Roman and Jewish worlds and of Christianity vary greatly in extent; there is separation into areas with Protestant or Catholic majorities; a further division is caused by legal concepts (Romano-Germanic or common law); there are countries with strong or weak colonial, maritime, or continental traditions; varying degrees of industrial, commercial, or financial specialization; unequal levels of development in different regions; traditions of trade with different parts of the world.

Amongst Europeans there is no end to the diversities rooted in history and consolidated by language.  The language problem is enormous, especially from the viewpoint of the accession of new member countries to the Community; however, apart from regional dialects to which some ethnic groups are particularly attached, the languages spoken fall into only two families, Germanic and Latin.[1]

Diversity is also found in the educational systems.  Young Europeans are educated in systems that differ in structure and content.  The same historical events are often explained quite differently.

The FISH survey on which this study is based shows that the European is extremely attached to these diversities.  He is so receptive to this awareness of difference that he finds it difficult to describe similar features and define his identity.

One of the most difficult problems to be overcome by the European Community is to profit from this diversity as a collective asset and, at the same time, to foster the awareness that a European identity really exists, transcending a mere association of economic interests.

Depending on background and geographical location, the attachment to diversity finds expression in a variety of ways.  To some, patriotism bound up with the concept of the nation is an essential value that must not under any circumstances be diminished by the formation of Europe.  To others, the basic cell to be respected is the Region, with its traditions, its language, and its specific culture.  Very few Europeans are prepared to be merged into a political entity that would eliminate a part, however small, of their diversity.  Feelings on this subject are generally expressed vehemently; intellectual reasoning is rapidly over-ruled by passion.[2]

This situation must be accepted as a fact, on the basis of which a political structure of a new pattern, not hierarchical, suited to complexity, confederal rather than supranational, may be devised for Europe.  These concepts are readily grasped by experts in systems analysis;[3] they realize that this view does not in any way diminish the importance and power of the confederal authority.  However, these "systems" ideas, although in no way intellectual but merely the outcome of experimental observation of living systems, are too new to be easily communicated.

---

[1] This fact indicates that machine translation (or machine-aided translation) by computer might not be unattainable.

[2] During the survey conducted by FISH this was one of the rare points on which those interviewed quickly lost their composure, showing the extent to which Europe would be harmed if it did not respect these attachments.

[3] See Fontanet, *Le Social et le Vivant*, Plon, 1977.

They will have to be put into practice by successive approaches; an evident
maturity will gradually emerge from this apprenticeship.  In the second part of
this study we propose that a start be made on this apprenticeship by
formulating and implementing a scientific and technical policy.

It may be said in passing that nothing, or almost nothing, has been done to help
young Europeans to realize that they are both specific to their country of origin
and specific to a Community.  For example, history is still taught as if the great
phenomena were national antagonisms, whereas the important facts are to be found
in the development of certain social classes, the content of education, the ideas
and legislation on employment, the status of women, and many other factors which
have developed simultaneously in Europe and have shaped the society of today.

Chapter 3

# ASPIRATIONS TO A NEW DEVELOPMENT

The "quality of life". What content could the European attribute to this
fashionable expression, rich in sentimental value but surprisingly difficult to
measure? In relation to what values, on the basis of what criteria, can the
quality of life be judged?

At a loss when faced with this problem, the best thing is once again to stick to the
facts and draw up a balance-sheet of what we know of or what we could do with:

(1) the time available to us in our lifetime;
(2) our communications facilities;
(3) our aspirations to growth;
(4) the new economic aims proposed to us;
(5) our working conditions and competitive situation.

### FREE TIME; THE HOPE FOR A NEW DEVELOPMENT[1]

Man's time is limited and he knows he has to die. Hence, is not his most precious
capital the time which each has at his disposal? Is not the most important
reflection in considering the criterion of the quality of life that which results
in the best possible use being made of the time granted us?

All civilizations have indulged in such speculation. For the ancients enjoyment
merged with "otium", free time, time which was noble, time for culture and
reflection, in contrast to "negotium", which was time for business. This concept
was widespread up to the advent of industrial Europe. It has constantly been in
conflict with the entrepreneurial need, the need to develop through action, the
need to participate in the production and business game. Indeed, otium was
reserved for the elite. It was the reward for power and wealth. In certain
degenerate forms such as idleness it became, especially in the Latin countries,
the sign that the person enjoying it belonged to the elite. The amount of free
time available to the few was but the result of arduous labour on the part of the
majority.

---

[1] This chapter is based on a study by Mr André Bonnet (IRIA, Rocquencourt) with the
help of documents and advice supplied by INED, Paris. The statistics are based on
observations for France, but the trends can be extended to the whole of Western
Europe.

A — HYPOTHESES ON LIFE-SPAN BETWEEN 1800 & 2000

| | | Observations made | | | | | Projections for 2000 | | |
|---|---|---|---|---|---|---|---|---|---|
| | | 1800 | 1850 | 1900 | 1950 | 1975 | A | B | C |
| (1) Life expectancy | years | 36 | 39 | 45 | 70 | 71 | 72 | 72 | 72 |
| | days | 13 000 | 14 000 | 16 000 | 25 500 | 25 900 | 26 300 | 26 300 | 26 300 |
| | hours | 315 000 | 340 000 | 390 000 | 610 000 | 620 000 | 630 000 | 630 000 | 630 000 |
| Breakdown per day (h) | Basic physiological time | $10\frac{1}{2}$ | $10\frac{1}{2}$ | $10\frac{1}{2}$ | $10\frac{1}{2}$ | $10\frac{1}{2}$ | $10\frac{1}{2}$ | $10\frac{1}{2}$ | $10\frac{1}{2}$ |
| | Travelling time | 1 | 1 | $1\frac{1}{2}$ | 2 | $2\frac{1}{2}$ | 2 | $1\frac{1}{2}$ | 1 |
| | Useful time | $12\frac{1}{2}$ | $12\frac{1}{2}$ | 12 | $11\frac{1}{2}$ | 11 | $11\frac{1}{2}$ | 12 | $12\frac{1}{2}$ |
| (2) Break- | Basic physiological time | 135 000 | 150 000 | 170 000 | 270 000 | 270 000 | 275 000 | 275 000 | 275 000 |
| (3) down | Travelling time | 15 000 | 15 000 | 25 000 | 50 000 | 65 000 | 50 000 | 40 000 | 25 000 |
| (4) through-out lifetime (h) | Gross free time | 165 000 | 175 000 | 195 000 | 290 000 | 285 000 | 305 000 | 315 000 | 330 000 |
| (5) Period of childhood | | 45 000 | 50 000 | 55 000 | 70 000 | 65 000 | 70 000 | 75 000 | 95 000 |
| (6)=(2)+(3)+(5) Total physiological time | | 195 000 | 215 000 | 250 000 | 390 000 | 400 000 | 375 000 | 390 000 | 395 000 |
| (7)=(4)+(5)=(1)-(6) Net free time | | 120 000 | 125 000 | 140 000 | 220 000 | 220 000 | 235 000 | 240 000 | 235 000 |
| Childhood duration | years | 10 | 11 | 13 | 16 | 17 | 17 | 17 | 21 |
| | days | 3 500 | 4 000 | 4 500 | 6 000 | 6 000 | 6 000 | 6 000 | 7 500 |
| | free hours | 45 000 | 50 000 | 55 000 | 70 000 | 65 000 | 70 000 | 75 000 | 95 000 |

B – HYPOTHESES ON WORKING LIFE BETWEEN 1800 & 2000

| | | Observations made | | | | | Projections for 2000 | | |
|---|---|---|---|---|---|---|---|---|---|
| | | 1800 | 1850 | 1900 | 1950 | 1975 | A | B | C |
| Active life | Starting age | 10 | 11 | 13 | 16 | 17 | 17 | 17 | 21 |
| | Retirement age | – | – | – | 65 | 63 | 57 | 63 | 64 |
| | Number of years | 26 | 28 | 32 | 49 | 46 | 40 | 46 | 43 |
| | Number of days | 9 500 | 10 000 | 11 500 | 18 000 | 17 000 | 14 500 | 16 500 | 15 500 |
| | Number of hours | 230 000 | 245 000 | 280 000 | 430 000 | 400 000 | 350 000 | 400 000 | 375 000 |
| Duration of work | hours per day | 12 | 12 | 10 | 9 | 8 | 8 | $6\frac{1}{2}$ | 8 |
| | day per year | 300 | 300 | 300 | 340 | 230 | 220 | 240 | 200 |
| | hours throughout life | 95 000 | 100 000 | 95 000 | 105 000 | 85 000 | 70 000 | 70 000 | 70 000 |
| Adult Life (work & retirement) | Working days | 7 800 | 8 200 | 9 500 | 12 000 | 10 500 | 9 000 | 11 000 | 8 500 |
| | Days of unemployment | 1 700 | 1 600 | 2 000 | 6 000 | 6 500 | 5 500 | 5 500 | 7 000 |
| | Days of retirement | 0 | 0 | 0 | 1 500 | 3 000 | 5 500 | 3 500 | 3 000 |
| Free hours | In working days | 4 000 | 4 000 | 20 000 | 30 000 | 30 000 | 30 000 | 60 000 | 37 500 |
| | In days of unemployment | 21 000 | 23 000 | 25 000 | 70 000 | 70 000 | 70 000 | 65 000 | 37 500 |
| | In days of retirement | 0 | 0 | 0 | 15 000 | 35 000 | 65 000 | 45 000 | 40 000 |

Hypotheses on work

C - RECAPITULATION OF RESULTS: TIME BUDGET BETWEEN 1800 & 2000

| | Observations made | | | | | Projections for 2000 | | |
|---|---|---|---|---|---|---|---|---|
| | 1800 | 1850 | 1900 | 1950 | 1975 | A | B | C |
| Physiological time | 135 000 | 150 000 | 170 000 | 270 000 | 270 000 | 275 000 | 275 000 | 275 000 |
| Travelling time | 15 000 | 15 000 | 25 000 | 50 000 | 65 000 | 50 000 | 40 000 | 25 000 |
| Free time in childhood | 45 000 | 50 000 | 55 000 | 70 000 | 65 000 | 70 000 | 75 000 | 95 000 |
| Working time | 95 000 | 100 000 | 95 000 | 105 000 | 85 000 | 70 000 | 70 000 | 70 000 |
| Adult free time | 25 000 | 25 000 | 45 000 | 115 000 | 135 000 | 165 000 | 170 000 | 165 000 |
| Total lifespan | 315 000 | 340 000 | 390 000 | 610 000 | 620 000 | 530 000 | 630 000 | 630 000 |
| % of time at work compared with {Whole life | 30 | 29 | 24 | 17 | 14 | 11 | 11 | 11 |
| {Active life | 41 | 41 | 34 | 24 | 21 | 20 | 18 | 19 |

How do matters stand today?  On the basis of average figures and extrapolations of
trends observed in the past fifty years, there have been very curious changes in
the time budget.

On average, and this is one of the most important consequences of technical progess
from 1800 until the present day in Europe, life expectancy has risen from 36 to 72
years, i.e. from 315 000 to 630 000 hours.  What has man done in qualitative terms
with this gift in the form of life?  In this field, too, the post-industrial
civilization brings profound changes.

Physiological time (Fig. 20) is that devoted to sleep, nutrition, hygiene, and a
minimum of repose in the waking state.  The specialists estimate that the average
for physiological time in a lifetime is 10½ hours per day.  In the whole of this
chapter we are talking of course, about averages per average individual (men and
women) who attain, but no more, the average expectation of life.  Taking into
account his longer total lifetime, the average individual would have spent 135 000
hours on physiological needs if he was born in 1800 and 270 000 hours if he was
born in 1950.

Travelling time in the urban environment (Fig. 22) has steadily increased during
the twentieth century.  The specialists attribute to this head, on average, one
hour per day up to about 1850, the figure rising to 1½ hours in 1900, 2 hours in
1950, and 2½ hours in 1975.  So the average man of 1975 will have wasted 65 000
hours of his life on travelling as opposed to 15 000 for a man in 1850 (in 71
years, of course, instead of 39).

"Prolonged childhood" has steadily increased in the form of successive extensions
of the period preceding initiation into working life.  On average it was ten years
in 1800 and seventeen years in 1975.  The principal change came between 1920 and
1950.  It was during this period that the school-leaving age was raised, on average,
to beyond the age of puberty.

The amount of time devoted to work remained, on the whole, stable from 1800 to
1960 at around 100 000 hours for a total lifetime, despite the doubling of the
average expectation of life.  If fell off markedly from the period 1955-1965,
reaching 85 000 hours in 1975.  Extrapolation of the variations recorded over the
last twenty years and observation of what is happening in the United States leads
to the assumption that, if the trend is not reversed, total working hours per
lifetime will have been reduced to 70 000 by the year 2000.  The factors which
serve to reduce working hours are known, i.e. a longer childhood, appearance of
retirement schemes, the holiday periods allotted, and a shorter working day.

The biggest change is taking place with regard to adult free time which includes,
in addition to the whole of the retirement period, minus physiological time, free
periods in the working life not devoted to sleep, nutrition, hygiene, and travel.
The "free time capital" granted to the average male and female population in 1800
was only 25 000 hours in a lifetime.  Between 1920 and 1975 a prodigious increase
was experienced and the figure rose from 45 000 hours to 135 000 hours.
Extrapolation to the year 2000 gives an estimate of close on 145 000, which could
rise to 170 000 if the waste in the form of travelling time were reduced from
2½ hours to 1 hour daily.

Even admitting that these figures have to be corrected by a fatigue coefficient
which has reportedly been on the increase since 1945 on account of the reduced
quality of the environment, making it necessary to increase by 1 hour per day the
amount of physiological time spent on indispensable additional rest, it is seen
that the major phenomenon of the last half century has been the increase in the
free time available to adults for things other than their elementary needs and
paid work.

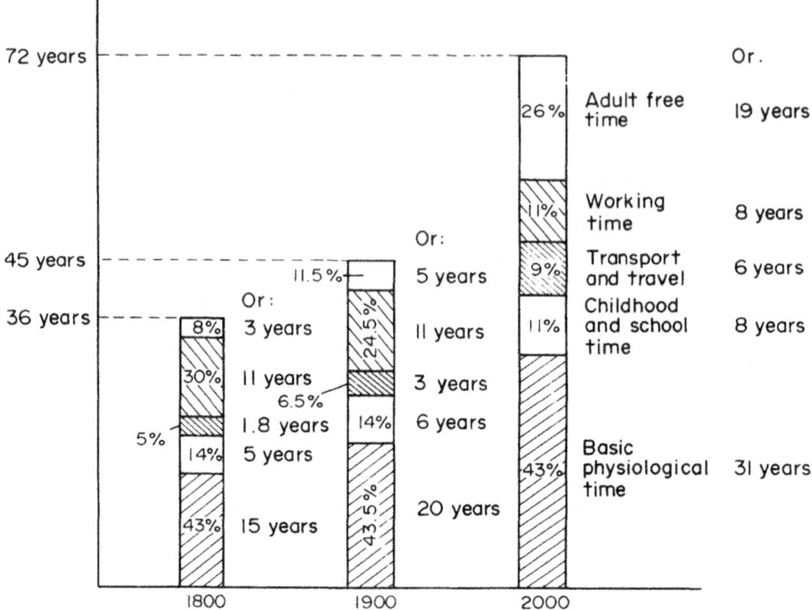


Fig. 20. Trend in the time budget of an average
individual – France. (Source: INED.)

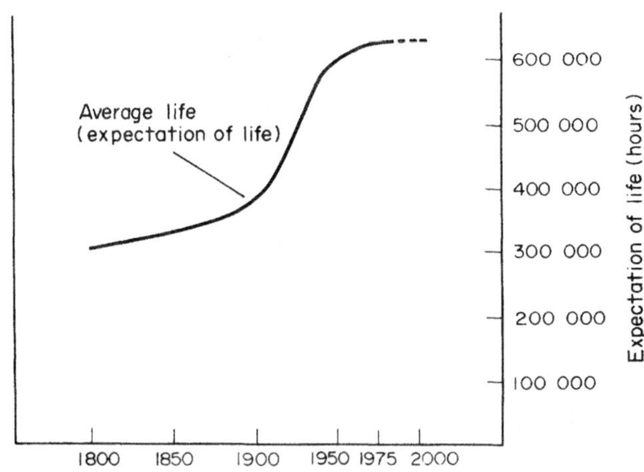


Fig. 21. Life expectancy.

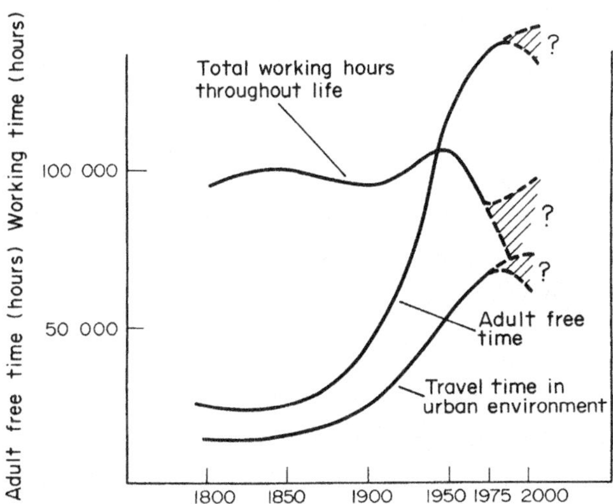


Fig. 22. Use of life.

These observations call for some comments if they are viewed in the context of an
attempt to improve the quality of life.

Is it a good thing to extend childhood very much beyond puberty?  Is not a late
start in social life and in accepting responsibility a serious factor in creating
lack of understanding and inability to adapt?  Would it not be better if the
acquisition of knowledge were not concentrated in a specialized educational and
training period lasting well beyond adolescence but, after a certain initial
education, took place in stages throughout the individual's life?

Should not one of the decisive steps forward in modern technology be to reduce
travel time?  Should this not be by means of an improvement of public transport,
and of town and country planning which gives priority to the reduction of
commuter distances by transporting not persons but information or products or
means of production?

Should adult free time continue to increase?  How will it be used?  Can the
proposal be made to people that they should spend their leisure time on increasing
their knowledge, their culture, the spiritual content of their lives, and
producing goods and services under conditions of free choice and with constraints
which cannot be other than foreign to paid employment?  Will man be happier if
he adds to his breadwinning activity a productive activity, a source of additional
income, taking an artisan, or artistic, or even utilitarian form such as cultivating
his garden or building his house?

In a Europe poor in resources by comparison with many of its world competitors and
yet desirous of procuring for its peoples a high standard of living, should the
total number of working hours in a lifetime continue to decrease?  At what price
in terms of automation in industry and increased productivity in services would
this be?  Can free-time activities, orientated towards the creation of products
and services, make up sufficiently for the reduced purchasing power which will
probably accompany the further reduction of paid work?  Does not the aspiration
towards a better life call for thought regarding free time, its breakdown over
the day, the week, the year, the lifetime?

Increased expectation of life (Fig. 21) is accompanied by the appearance of a
category of very aged persons which was negligible in terms of numbers in the recent
past, but it is now close to 3.7% of the population in 1975 and will reach more
than 5.5% in the year 2000.  Should society devote its attention now to the problems
they will cause?

*  *  *

These questions probably help us to a better understanding of the problem of the
quality of life.  If the trends towards reduced working hours could be prolonged
and if a consensus could be reached to the effect that most of an adult's available
time would be spent on productive activities or activities in the direction of
individual further training, an immense field for growth of a new type would be
opened up in handicrafts, the minor and major arts and the further acquisition of
knowledge on the part of adults in all disciplines.

Free-time activities do not consume much energy or raw materials.  They are therefore
compatible with Europe's relative poverty as regards primary resources.  Of course,
there could be no question of a return to the processes in use in cottage industry,
activities typical of the pre-industrial centuries.  Production at home would be fed
at source by an industry specializing in prefabricated products or specialist tools,
but it is to be supposed that a not inconsiderable added value would be supplied by
the labour contributed by users themselves or by their relatives and friends.

If the reader were to meditate on the following list he could perhaps add some lines
to it?  What would he like to do if he had the time and the facilities?  What new or
further knowledge would he like to acquire?  What special subject would he like to
add to the obviously incomplete list given here of the various sources of production
at home or self-study?  In these free-time activities, how could he introduce the
sort of conviviality which is not feasible in industry, trade, or administration
because of productivity requirements?

*Free-time activities (unpaid)*

(1)  Games and entertainment

Individual, team, and parlour games and pastimes.  Family celebrations, office or
group outings or parties, etc.  Keeping up local traditions.

(2)  Arts taking the form of personal expression

Writing of all kinds (novels, poetry, essays, memoirs, etc.).  The plastic arts
(drawing, painting, sculpture, engraving, etc.).  Scenic arts (music, singing,
choral singing, dancing, theatre, puppets and marionettes, etc.).  Audiovisual
arts (photography, filming, sound recording, etc.).  The applied arts (decoration,
cabinet-making, ceramics, tapestry, etc.).  The culinary art (gastronomy).

(3)  The further acquisition of personal knowledge

Lifelong education.  Vocational retraining or further training.  The acquisition
of social and economic culture (knowledge of society).  Travel for cultural
purposes.  Participation in archaeological, ethnological, ecological, or
historical research.  Learning a personal activity (art, craft, sport).
Preparation for political or social work.  Learning a foreign language.

**(4) <u>Contribution towards the acquisition of knowledge</u>**

Participation in study circles, clubs, scientific societies, etc. Scientific
research without costly equipment.  Voluntary work in non-profit-making
organizations Educational activities (courses, lectures, demonstrations).

**(5) <u>Civic, religious, and spiritual activities</u>**

Political or trade union work.  Practice of a religion; pilgrimages; study of
religions.  Philosophical meditation; spiritualism; parapsychology.  Participation
in psycho-sociological survey.

**(6) <u>Aid to the developing world</u>**

Participation in study groups, collection of aid funds.  Welcoming activities for
nationals of developing countries.  Participation in friendly missions,
technical assistance, missions, etc.

**(7) <u>Social and family life</u>**

Participation in local life (organizations, local authorities, regions).  First
aid.  Child care and education.  Care of various kinds (the sick, the elderly,
the handicapped).  Assistance in the rehabilitation of the injured.  Participation
in the management of facilities for the young, the old, the handicapped, etc.
Prison welfare work.  Assistance with leisure activities for the elderly.

**(8) <u>Sports</u>**

Gymnastics and open-air activities (swimming, walking, climbing).  Leisure sports
(fishing, shooting, riding, boating, etc.).  Competitive sports.  Team sports.
Rambling, cycling, trekking on horseback, etc.

**(9) <u>Crafts and handwork</u>**

Do-it-yourself building, house improvements, furniture-making, construction of
heavy sports equipment (aircraft, boats).  Carpetmaking, tapestry, basketwork,
pottery, ceramics, enamelling, mosaics, bookbinding, carpentry, cabinet-making,
marquetry, etc. Weaving, dressmaking, lacemaking, knitting; making and
restoration of traditional costumes, fancy dress, etc.  Working of wood, leather,
ivory, plastics, precious metals; making of costume jewellery.  Toymaking.
Horticulture, gardening, poultry rearing, stockbreeding, etc., for one's own
use.  Do-it-yourself activities and maintenance work (electricity, plumbing,
wallpapering, car, etc.).  Restoration of furniture, curios, clocks, frames,
pictures, musical instruments, etc.  Restoration of ancient monuments and
historical sites, participation in archaeological digs.  Assembly from kits of
radios, hi-fi systems, TV sets, radionavigation equipment, sound and light
equipment, burglar alarm systems, calculators, etc.

The provision of a development framework for free-time activities in order to
channel them towards production in the home would more or less amount to guiding
and legalizing what is today known as "moonlighting", which appears to meet a
general desire on the part of the peoples of Eastern Europe and also in some
members of the Community of the Nine.  This work on the side partly accounts for
the resistance to loss of earnings where strikes and unemployment greatly reduce
official output.  The success of this type of work also shows that there is a
strong movement in favour of a productive activity even when circumstances favour
idleness and passive free-time activities, provided there is a serious economic
incentive combined with a feeling of liberty.

*　*　*

The above statistics on the time budget are mean figures, which implies that
particular cases are not really taken into account.  The organization of free
time may be considered to vary throughout a person's life:  part-time working,
early retirement, sabbatical years for vocational training, reduction in weekly
working hours.  *A la carte* arrangements of this kind would probably bring back
to work a large proportion of the non-working female population, which would
meet a deep-felt desire.  The working population would then be greater in number
but each individual would work shorter hours.

It is necessary to re-emphasize, however, the major difficulty already pointed
out.  The hard industrial and commercial competition that Europe will have to
face, combined with its natural economic handicaps, will not allow existing
standards of living to be preserved without a considerable effort on the part
of its peoples.  The extrapolation of the tendency towards a reduction in
working hours that emerges from today's statistics would be dangerously utopian if
the following two conditions are not met in the future:

(1)   substantial further improvement in the productivity of industry and services;

(2)   effective contribution made by free-time production to economic wealth by
      the provision of products of arts and crafts that are useful in everyday
      life or serve the purpose of cultural enrichment.

*   *   *

There are other reasons for devoting serious thought to the question of free time:

(1)   Modern society lacks facilities for communal life such as were provided by
      the village or parish in earlier times, and consequently a greater burden in
      the care of the individual falls on the home and on the place of work.
      Inevitably, there will be conflicts in the relations between people.  This
      emotional overburdening at home and at the place of work is a hazard that
      may disrupt them both.

(2)   The unsuitability of higher education to meet real requirements, rampant
      unemployment, and illegal immigration are all factors encouraging drop-outs;
      post-industrial society in its present form creates drop-outs.  Might not
      the idea of free-time activities extended to the production of cultural
      goods or official recognition of what is today "moonlighting" help to
      improve this very dangerous situation?

*   *   *

The risk of an increase in passivity, or still worse in alcoholism, drug-taking,
and the crime rate must be set against these optimistic ideas on free-time
activities.  What resources can the human sciences employ to generate a new
social consensus, new systems of values that will enable us to overcome the
difficulties while respecting individual freedom?

These reflections on the time budget give rise to many questions to which there
is still no answer for lack of sufficient information, but we must act quickly
as everything points to the fact that in the area of working hours and the
development of production in the home, prudence calls for a spirit of enterprise
and the taking of a calculated risk.

## INFORMATION ACTIVITIES:  FREE SCOPE FOR NEW GROWTH GROWTH[1]

Up to the nineteenth century civilization was mainly rural.  Agriculture occupied
the greater part of the working population.  Then came industry.  It developed
at first without noticeably changing agricultural work — the working population
had become more numerous as a result of the flareup in the birth rate.  Then competition started and this obliged farmers to mechanize in order to step up consider
ably their productivity per man employed.  The rural population dwindled.  The
developed countries became urban.  The proportion of persons working in industry
increased greatly at the expense of all other forms of activity.  Society became
rapidly more complex and this made necessary the creation of services and information processing of all kinds, i.e. administrative, financial, commercial, and
technical.  It called into being new specialities in the acquisition of knowledge
and its communication.

More recently the productivity of agriculture and industry has been making prodigious progress, freeing jobs and increasing complexity.  Labour is shifting to
the services sectors.  For the United States, which is ahead of Europe in this
process, the percentage curves for the working population on an annual basis are
shown in Fig. 23.  A distinction is drawn between three types of activity:
agriculture, industry, and services.

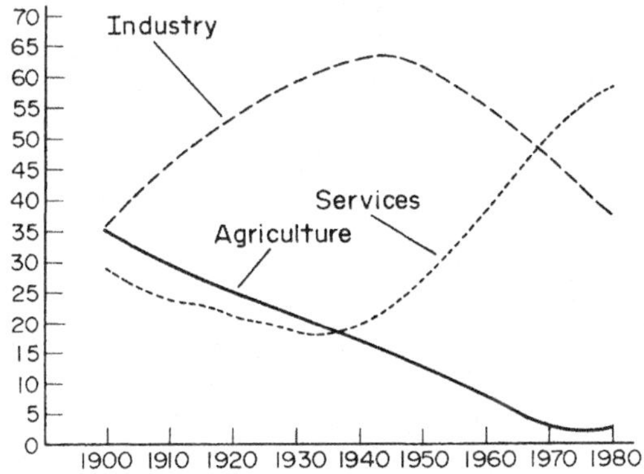


Figure 23.  Working population of the United States:
percentage breakdown into three sectors.

Edwin Parker proposes, following upon Mark Porat, to hive off from each category
of workers above those who specialize in the collection, processing, and spreading of information.[2]  He distinguishes four sectors: the three conventional
sectors excluding all operations on information plus a fourth information sector.
The curves showing the proportion of employed persons since 1900 for the United
States are given in Fig.24.  They show a real change in the nature of employment
which became apparent in the period 1955-1965.

---

[1] This chapter is largely based on the lecture given by Professor Edwin Parker
at the OECD on 4 February 1975 (OECD Document DSTI/CUG/75,1, 27 January 1975).

[2] "Information" here is used in a fairly wide sense of statistical data, general
knowledge, news information, thought, etc.

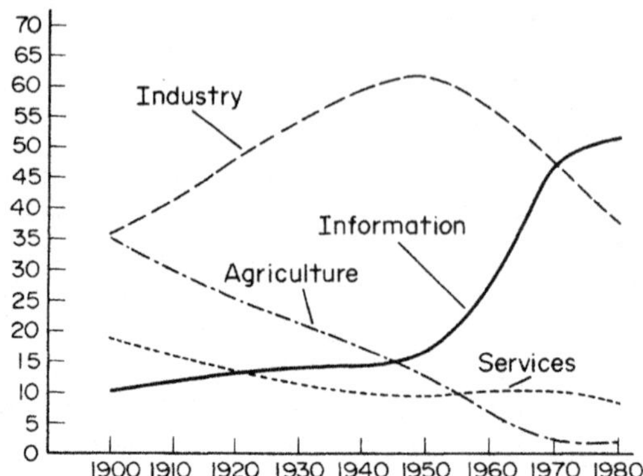


Figure 24.   Working population of the United States:
percentage breakdown into four sectors.

One can see that, in the decade 1970-80, those actively employed in information
activities will exceed those employed in work exclusively of an agricultural,
industrial, or service nature.

As far as these phenomena are concerned it is true that the trend in Western
Europe is behind that in the United States, but events are following the same
course.  The delay may be estimated at from five to ten years.  It is legitimate
to say, therefore, that, by the nature of things, if the trend is not nipped in
the bud by an excessively severe economic crisis, it will impel us in the direction of an economy where the main proportion of activities will be linked with
information.  For Europe, if this comes to pass, the observation is fundamental,
for the collection, processing, and dissemination of information does not consume much energy or raw materials but it does require numerous brains of a high
quality.

There follows a list, obviously not exhaustive, of the main specialities coming
under the head "information activities".  It is obvious that most of these fit
in remarkably well with the resources and aspirations of European culture, as
already analysed previously.  At professional level, they overlap some of the
free-time activities studied in the previous chapter and in some cases provide
technical or logistic support for them.

*Principal Information Activities*

(1)  <u>Acquisition, storage, and transmission of information</u>

Scientific and technical research.
Documentation, libraries, data banks.
Training, education, communication of cultural, scientific, and technical
knowledge in all its forms (further training, language laboratories, etc.).
Learning a craft.
Intellectual and artistic creation:  literature, painting, sculpture, music,
etc.
Entertainment (theatre, circus, cinema, etc.).
Museums, exhibitions, trade fairs, etc.
Audiovisual activities (broadcasting, television, audio and video reproduction,
etc.).

Written and spoken press, advertising.

## (2) Acquisition and processing of information

Applied mathematics.
Forward studies, planning, market research.
Statistical observations, modelling (in economics, town and country planning, transport, etc.).
Standardization.
Project studies.
Technological innovations (development, demonstration, promotion, patents).
Exercise of a political or trade union responsibility; management of a collective; conduct of an administration or a public service; management of an enterprise (executive and supervisory staff).
Drawing up laws and regulations, justice.
Medicine, health.
Meteorology.

## (3) Services based on information

Printing.
Postal and telecommunications services.
Banks, financial services, stock exchange.
Social security.
Insurance.
Organization of spare time.
Assistance to developing countries, transfer of technology.

## (4) Support industries (Specialist equipment and supplies)

Electronic components.
Printing presses, tele-typesetting.
Reprography, telereproduction, photography, sound recording.
Telephone, telex, videophone, teleconferencing, electronic mail.
Electronics: broadcasting, and television (transmitters, studios, receivers), radio links, tape recorders, hi-fi, sound proofing, etc.
Computers, teleprocessing, networks, microcomputers.
Automation (sensors, microprocessors, etc.).
Scientific instruments.
Information media (photographs, films, discs, memory, etc.).
Software industry; computer service activities.

## THE CALL FOR A NEW DEVELOPMENT

It is necessary to have experienced the rapid growth of an industry to realize what enthusiasm is generated in a management team by an increase in turnover, the launching of new products, the employment of an ever-increasing work force, or the conquest of new markets. Growth cures or masks all diseases; it leads to full employment, gives everyone a feeling that progress is being made, even if accompanied by inequalities, overcomes financial obstacles, gives rise to enormous productivity gains, and, even more important, gives man the feeling that he is useful, that he is participating in a major undertaking, and that the effort called for is something of a pioneering venture. The post-war generation experienced this exaltation of growth to the highest degree in the industrialized countries up to the end of the 1960s. The intensification of the drive for material power was the continuation of a long progression.

Observations over the past centuries of history show that growth is a typical
phenomenon of the human race as the ravages of catastrophes, famine, epidemics,
and wars have always been rapidly repaired.  Globally, all curves are ascending:
population, capacity and number of production tools, capacity of economic production, knowledge of nature and of man, communications facilities in the way of
transport and telecommunications, acquisitions of a cultural, aesthetic, and
philosophical nature, destructive power of arms, etc.  This overlapping of growth
phenomena is known collectively as development.  In the rich countries the rate
of development has speeded up surprisingly since the Second World War but in an
imbalanced manner; technological progress has moved apace, breaking away to some
extent from a cultural evolution characterized by the continuance of concepts
derived from the past.  Growth has no longer taken place primarily in the intellectual sphere, but in the field of material consumption, not without bringing in
its wake an increase in all the phenomena of complexity and interdependence.

The Club of Rome[1] must be given great credit for having dared to announce that
we were approaching the limits to exponential growth in the consumption of scarce
materials and energy, that the world had finite dimensions, and that we were coming very close to those limits.  And even if the mathematical model used to
demonstrate these findings was marred by undoubted errors, much of what the message announced was true.  But this warning disrupted the comfort guaranteed by
the growth mystique;  contrary to its author's intentions, it was interpreted
as reflecting a deep pessimism.

Recent years have helped to define the limits of the dangers.[2]  The risks of
shortages of primary resources are small in the short term, with the exception
of two — unfortunately essential — items:  petroleum and, for the peoples of the
tropical and equatorial belts, food.  Experts agree that pollution could be contained within acceptable limits if 2 to 2.5% of the gross national product were
devoted to its control.  However, for some products such as phosphates, mercury,
silver, chromium, and a few others, the scarcity of easily mined high-grade ores
could lead to serious price increases.  Nothing fundamentally tragic for humanity
results from this analysis;  the younger generation must not view the future as
inevitably destined to produce apocalyptic catastrophes, but the warning is a
serious one.  In the present state of technology and the possibilities for capital
formation, the spread of the American consumption pattern to the rest of the
world would lead to enormous wastage.  And yet the American successes are taken
today as a yardstick and aim by almost all governments implementing a deliberate
development policy.

Europe must give serious consideration to the question of growth as the finite
nature of the world and its interdependence have a profound and immediate meaning
for it:  they recall its cruel shortage of primary products and its dependence
on suppliers over which, as we have already seen, it no longer has any political
control.

In recent years thinking in intellectual circles has appeared to fluctuate
between two extreme solutions, both of which are impractical:  the restimulation
at all costs of consumption growth as experienced for rather more than twenty
years after the Second World War and, at the opposite pole, zero growth.

---

[1] The founder of which, Dr. Aurelio Peccei, is a member of CERD.

[2] See in particular the symposium organized by FISH on the terrors of the
year 2000 in September 1975, in Jouy-en-Josas.

*Rejection of Zero Growth*

The choice is shown fairly clearly in Fig. 25.  Zero growth would mean that the
curve representing economic activity would lie between curves I and I bis.  For
the first time in the history of humanity, apart from temporary short-lived economic crises, the second-order derivative of the function economic activity would
take on negative values and would remain negative.  It is by no means certain
that we would be able to handle a situation of this kind, demanding reflexes
so different from all that we have acquired in recent decades.  And it is difficult to imagine what a society compelled to embark on deceleration would be
like.  In all probability, an economy of this kind would necessitate the freezing of existing structures and a great rigidity of society.  Instead of continuously encouraging individual opportunity and rewarding a spirit of enterprise,
it would enclose people in fixed organizations in which advancement would depend
solely on seniority.  Society would probably evolve by moving towards crystal-
lization into a structure with no mobility.  The absence of reward would not
encourage effort.  The freedom of individuals and groups would be greatly reduced
by the fragility of the static balance, whereas the adaptability of dynamic
growth balances easily absorbs errors and flourishes on successful endeavours.

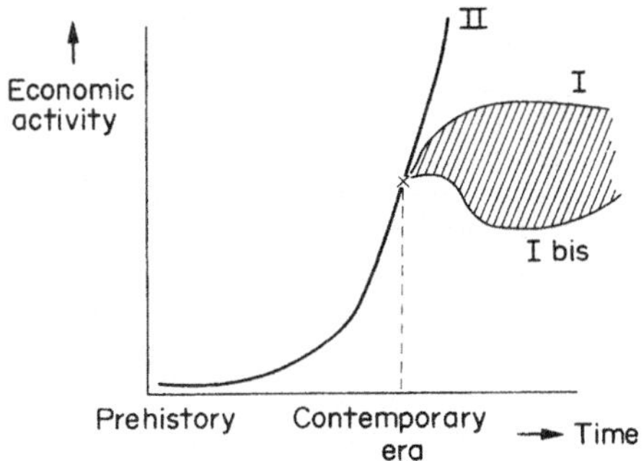


Fig. 25.  Growth in economic activity.

Would this sedimented society be based on a system of egalitarian distribution,
with neither wealth nor poverty, or on a caste organization established by
compulsion?  Either of these formulae could be adopted, depending on the country
or circumstances.

Zero growth means seeking a static equilibrium.  Engineers know that there can
be balance only in a combination of dynamic forces, and biologists have proved
that life is essentially movement, happening, revival, conflict, surprise.  But
static equilibrium is attractive to many minds;  it appears endowed with capacity
for refinement that would enable perfection to be attained;  it would guarantee
everyone security, would eliminate conflicts of interests because these would
become pointless.  It is the temptation of nothingness, the fascination with
death to which man is so sensitive.  It is to be expected, therefore, that we
will encounter in the future the manifestation of dreams rooted in the past, myths
of the return to a rural civilization within closed geographical frontiers.  As
a result, the tendency towards authoritarian organizations in which the State,
with all powers concentrated in it, would decide on the happiness of its citizens,
is a likely outcome, as the positive constraints emerging from a collective desire for development can be replaced only by political constraints.

*Discovery of the Sources of a New Development*

The European needs to develop through action, as we have said: he is unable
to divorce thought from action; he is not prepared to merge his personality in
the mass.  A programme of aims must therefore be proposed to him.  However,
this programme must turn its back on excessive consumption of energy and raw materials, must avoid waste and attacks on the environment.  It is necessary to
invent a new development.  Only by proving that it is feasible to enter a phase
of development of an entirely new nature will it be possible to create the
necessary willingness, give new impetus to the spirit of enterprise, and release
the forces of change.  This will overcome opposition to further productivity
increases at work because the fear of unemployment will be gone.  The announce-
ment that a new growth model is in sight is the only way of giving youth faith in
its future and persuading it to sacrifice the time and effort necessary to achieve
success.  This is an essential condition for the rehabilitation of the concept
of progress.

* * *

All these ideas are in the air; they have been expressed in many countries and
in a variety of environments but have not brought matters much further forward.
One would have to be extremely vain to claim to have found solutions; at the
most one may hope to foster emulation, to set brains working, and to arouse cri-
ticism from which constructive proposals will emerge.

In the course of the two previous chapters, two fields have emerged as possible
growth areas which have no scarcity constraints and are in line with the drive
towards psychism typical of evolution.  Man fulfils himself by making a judicious use of free time; he increases his ability to communicate with his fellows
and with his environment by means of information activities.  These are the new
poles of growth but, for economists, it will be clear that these aims will not be
sufficient to trigger off a movement towards a Second Renaissance.  Two considerations underlie this scepticism:

(1)  Never in history have men succeeded in uniting their economic forces for the
     sole aim of improving their lot, the drive towards comfort and pleasure aris-
     ing from the cultivation of the senses and of the mind is not a rallying
     force; perhaps men are still too primitive and too unequally cultivated?

(2)  Assuming that these aims are acknowledged as desirable by a large majority
     of the population, resources to finance the transitional phase are inadequate.
     Europe does not create enough capital to pay for the new communal and indi-
     vidual investments; it does not derive sufficient surpluses from its economic
     activities to absorb the costs of the social and commercial imbalances gener-
     ated by a transition from a type of civilization essentially based on the
     production of material goods to a civilization of communication and of the
     mind.  That is why it is urgent to devise a third pole of growth of a more
     familiar kind, more in line with industrial investment and the technical re-
     sources available in the industrialized countries.  If this third pole of
     growth were indeed to be able to continue to foster productivity increases,
     and if these increases could generate sufficient economic surpluses, then the
     financing of the transition towards a Renaissance would no longer be an in-
     surmountable obstacle.  Prospects of a society more richly endowed with
     humanity would no longer be merely Utopian but would become a project.

Asking this question is tantamount to pondering on the market: where are the real needs? The reply is clear: the real needs for an increase in consumption may be found amongst the 2,000 million (who will shortly number 4,000 million) who are today deprived.

## EURO-SOUTH SOLIDARITY

It is not possible in a short chapter to deal with the vast question of the economic takeoff of the poor countries. However, it is impossible to speak of the economic future of Europe without referring to the enormous development possibilities, for the Community itself, which would be opened up by financing, if this were possible, the equipment in capital goods of the countries in the process of industrialization. A new growth extending over thirty to fifty years would be open to European companies. If this process could be triggered off, it would bring about a fantastic economic spin-off in the wake of which the evolution towards a new quality of life, which appears Utopian today, could become reality.

Although the order of magnitude of the circumstances and the complications involved are not comparable, the situation has some similarities with the picking up of the economic pieces of Europe and Japan after 1945, in particular thanks to the loans under the Marshall Plan and their extensions. This act of solidarity was not only generous. Was it in fact dictated by generosity? However, it was intelligent and in recompense, directly and indirectly, the United States has received as a contribution to its own development several times the amount of the initial financing.

Many ideas are in the air today about the remission of the debts now crippling the economy of the developing countries, Sweden having just provided one example. There is talk of the financing of joint development, as proposed by Mr. P. Cheysson in particular; of a standstill on armaments which today scandalously consume an appreciable proportion of the economic growth potential of the poor peoples. Perhaps we are not far from the point of no return, the transition from theory to action?

Even if no great policy can be put on foot, however, the purchasing power derived from the increase in prices to suppliers of products that have become scarce (petroleum, gas, ores, tropical food products, etc.) is starting, at least in certain geographical areas, to stimulate takeoff. Europe is directly involved; it must see that its own development shares in this trend.

## COMPETITIVENESS AND SIZE

Western man is not characterized by frugality.[1] He is used to an abundance of material goods, to civilized comfort, to a car, and a colour television set. His diet is rich. He aspires to a second house or dreams of travelling to far-off places or going to the theatre or the museum or practising winter or seaside sports. He considers it legitimate to enjoy a long retirement, calm and well financed. He intends to preserve his health by means of curative and preventive treatment. He wishes his children to study for prolonged periods of time and, if

---

[1] With the exception, however, of a poor sub-proletariat benefiting but slightly from the advantages gained through technological progress, the existence of which is a permanent insult to the ethics on which the industrialized society would like to pride itself.

possible, to prepare for a career associated with non-manual tasks.  He wishes to
be guaranteed against loss of his job and against the consequences of illness.
If he does not have the benefit of these he feels deprived.

This situation is due to the progress made in the productivity of the mass manufacture or consumer society, also known as the society of abundance.  No government would dare to let it be understood that the objective is progressively to
abandon what has been achieved because Europe's position in the world has changed
so much that it is no longer able to grant its populations such favourable treatment.  However, let it be said to all those with no political attachment who are
devoting their attention to Europe's fate that all these current advantages are
extremely fragile and if they are to continue it can only be as a result of considerable efforts consented to by the population at all levels.  The balance-sheet
for Europe drawn up in the first part of this document is sufficiently explicit
to demonstrate how remote the chances are.  To be without any great political
power in a world characterized by a pitiless competitive struggle in the division
of labour, when one has no natural advantage save the quality of one's population,
does not make for commercial victories without an iron will and a broad social
consensus.

Europe must therefore continue to increase the productivity of its agriculture,
industry, and services in order to remain competitive.  It must also pursue its
effort in the direction of innovation in order to surprise its competitors by its
technical progress.  On this basis it will be able to continue exporting in order
to import energy and the other raw materials which are indispensable for its economic life.  However, the inhabitants of Western Europe must understand that every
year they will feel weighing more heavily upon their shoulders the burden of a
finite world, the scarcity of primary resources, the need to eliminate pollution.
In this effort, one of the main weapons will be innovation.

Innovation is the result of scientific research.  In this field it would be well
for Western Europe to choose its ground, in keeping with its specific resources,
its dimensions, and its own cultural genius.  We shall come back to this later.


### Productivity and Size

As far as productivity is concerned, the European Community must not forget the
implacable rules of the economy of scale.  When a product or a service can be
obtained by means of a mass-production process its cost price is in inverse ratio
to the quantity manufactured.  Figure 26 explains this Law.

Depending on how far artisan manufacture is developed or whether it is partly
or completely automated, the price of a specific product will decrease in accordance with a curve which looks like a hyperbola; the more sophisticated the manufacturing methods, the lower its asymptote, the ultimate price.  However, it should
be noted that in the case of short-run production rustic methods are the most eco-
nomical. In other words, not only is the price a function of market breadth but
also of the degree of sophistication characterizing production.

Most of the good and evil attributed to the consumer society is contained in
these curves, namely:

(1)  the considerable gains in terms of productivity and the spectacular lowering
     of sales prices;

(2)  the strong trend towards giant production units;

(3)   the success of the multinationals;

(4)   the pressure exerted on customers for increased consumption of products;

(5)   the growth of development and engineering expenditure;

(6)   the tendency towards a lowering of quality standards, especially at the expense of durability;

(7)   the difficulties experienced in transferring rich country technology to poor countries whose domestic market is too small unless they are made into exporters with a controlled foreign market (neo-colonialism);

(8)   inflationary tendencies which are partly derived from the fact that the industrial cost price is deceptively low because the giant production units pass on to society a large proportion of their overheads in the form of the elimination of nuisances, transport of personnel, giant-scale urbanization, etc.;

(9)   the difficulties experienced in regaining a market once lost for a certain time (electronic components, microprocessors, computers, quartz watches, etc.).

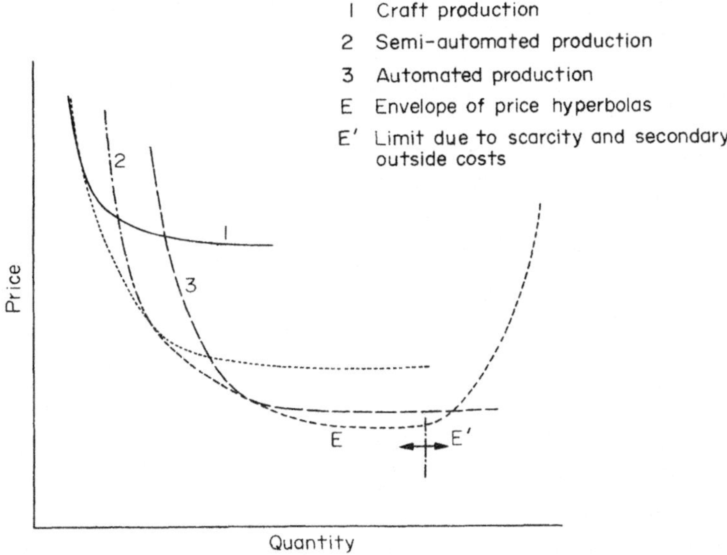


Fig.26.  Mass production:  price-quantity function.

European industry is finding itself confronted with the implacable nature of this law which lays down that price decreases with the quantity manufactured and, unlike Japan which exploits all aspects of this remarkably well, it seems that the Community has a special propensity for falling into all the traps which accompany these phenomena.

In the first place, the Community takes no advantage of its size.  As a result of the direct or indirect control of the Member States over the main markets for

innovation such as national defence, health, telecommunications, education,
and the nuclear field, despite the theoretical freedom of movement for products,
national producers are given preference.  Thus what should constitute a single
market of 250 million consumers with a high purchasing power and a much richer
potential than Japan, almost as powerful indeed as the US market, is broken up
into separate national units.  However, although Europe is singularly unable to
derive an advantage from economies of scale it is not able to combat the dangers
thereof.  In particular, it is allowing itself to be dragged into the production
of goods with a short lifetime intended to be thrown away and not repaired; this
is in no way suited to its relative poverty in terms of energy and raw materials.[1]
No great efforts have been made to recycle waste.

In addition the Community has paid little attention to real cost prices.  This
must include both direct industrial costs and expenditure outside the undertaking
necessitated by the growth of high-density urbanization and the vulnerability of
the environment.

It is therefore not only in the field of customs that Europe, as is often said, is
borrowing from the United States the worst without taking the best.  It is also
true, in many respects, in the economic field.

From this glance at the economy of scale it must be deduced that Europe will not
maintain the purchasing power of its peoples in international competition if it
does not accept certain constraints which are imposed by the laws governing price
formation in mass production and particularly if it does not unify its main markets for the products of innovation.  With the help of technical standardization
and regulations, it should introduce new constraints in line with its special characteristics and should, among other things, steer its industry in the direction
of products with a long lifetime, which can be repaired and are not rapidly consumed,
and products designed so that the scarce raw materials embodied in them can easily
be recycled.  Its efforts should also be directed towards a closer approach to true
prices so that the protection of its social and ecological balance, which is particularly delicate, can be allowed for.

*The Concentration of the Large European Undertaking and the Challenge
of Multinational Business*

As many observers have commented, the need to become competitive by increasing
the scale of operations and the shock caused by the Treaty of Rome have led
not to intra-European concentration, as might have been expected, but to mergers
between rival groups within member countries.  Examples probably come to mind readily in Germany, France, and the United Kingdom, where the chemical, electrical engineering, steel, and other industries set to work to form powerful national groups.
A similar course was adopted in Italy by the State financial trusts IMI and ENI.
Instead of market interpenetration, the formation of these oligopolies has resulted
in a strengthening of specific national features - exactly the opposite of what
should have been done to offset the power of the major American or Japanese industrial groups.  As a result, with, of course, a few exceptions, real transnation-
ality within the Common Market has been achieved only by European companies based
in the small countries (Netherlands, Sweden, Switzerland) on the strength of the
momentum acquired from their long-established location and by the ventures of
American companies.  The latter, inward-looking right up to 1957, were suddenly
stimulated by the rebirth of Europe to engage in ventures outside their traditional
home markets.

---

[1] This statement can be disputed; it is not true of all industrial products. This
controversy shows the need for serious studies on this subject.

The question that arises concerns not only industrial competitiveness but also
the dissemination of innovation.  The development of modern techniques is so expensive that the costs of research and development, demonstration, and sales promotion must be recovered on the markets of several major consumer countries simultaneously.  Here Europe is faced with a problem that has not yet been properly solved and that comes under both the industrial and the scientific and technical policy.

These reflections on the economy of scale provide an opportunity to illustrate a
contradiction.  To remain competitive, which means merely to survive, European
industry is forced to accept certain trends towards mammoth industry.  And yet,
as we have seen, mammoth production units are not at all in keeping with the
geographical characteristics of Western Europe and all that we have said of social
aspirations and cultural forces does not indicate any natural predisposition to
succeed in large concentrations.  The difficulty cannot be resolved by an all-or-
nothing choice.  There are fields where Europe is obliged to accept mammothism.
It can succeed if it wants to.  But the ideal would be to develop consumption
models that are suited to production in small units; in this field the Common
Market countries would probably be winners.  This offers a promising line for
research.


## WORK AND QUALITY OF LIFE

The ideology that has accompanied the work results from a conflict between two
concepts which sometimes coexist in the mind of one and the same person.  Work is
a curse — the sign of man's imperfection.  Work is a matter of dignity; man
achieves greatness in work.  Depending on where one stands geographically or socially, or the point in history one considers, one of these concepts takes prece-
dence over the other.  Nevertheless, work and effort have been regarded as basic
values of society throughout the industrialization period, and the concept of dignity through work is still widely prevalent.  Only the youth of today question this,
but their challenge is perhaps directed not so much at the principle in itself
as at the fact that attractive aims on which to focus their effort are no longer
offered to them.  It is fair to say that as long as the tasks involved in paid employment are, rightly or wrongly, regarded as repugnant by a considerable proportion
of the working population, there will be no satisfactory reconciliation between the
citizen and society.

The place of work is generally the "undertaking", viewed in the broadest sense
of the word.  This is a human community specializing in a group of specific tasks.
The undertaking may be an industrial and commercial company, but it may also be a
non-profit-making venture; it may also be the administration of a university, hos-
pital, or local authority.

We shall not here go into the conflicts of ideas on profit and its distribution
except to warn against the risks of misleading simplification.  The complexity of
the modern world does not allow such simplification; in particular, it is wrong
to regard profit as the sole aim of the industrial and commercial company.  As for
power, because of the complexity of the situation, it depends far more on information than on ownership.

Here we are interested in the quality of life of the man at work: it is recognized
that this must be improved both in the public services and in private companies
and, according to our observations, in both free and planned economies.

Considerable research into means of further progress here is currently underway
in all the Community countries.  Because this is done, everywhere, in a state of
tension and sometimes in the face of trade union and employer antagonism, and because it is accompanied here and there by apparent anarchy due to confusion as to
who is competent, it tends to be disturbing.  It would be more constructive to
ensure that all the questions start off by being well formulated and that individual solutions get as far as experimentation.

On the basis of what has already been achieved until today it is possible to
classify the problems into two broad categories:

(1)   understanding of an undertaking's objectives and constraints and association
      with decision-making machinery;

(2)   elimination of distasteful tasks.


*Understanding of the Undertaking and Participation in Decisions*

Although the concepts that follow correspond to a level of information which is
not very widespread, men of goodwill are beginning to agree on concepts for
industrial, commercial, or service undertakings which differ noticeably from
the image inherited from the past.

The undertaking must be recognized as the principal centre for the production of
wealth.  Its aims are complex and essentially economic and social.  The undertaking processes raw materials or information in order to procure goods for its
customers, purchasing power for its employees, a return on capital for its share-
holders and creditors, while swelling the resources of the community via a number
of tax levies to which it is subject.  The undertaking also provides its suppliers
with a living.  Profit is an indicator of good adaptation; a brake if it is negative and an accelerator if positive.

The undertaking is, by its very essence, a place of tensions and conflicts.  The
interests of its personnel, which are to be as well paid as possible with working conditions, guarantees, and holidays which are likewise as good as possible;
the interests of its customers who seek the best quality at the lowest price; and
the interests of its creditors and of its suppliers, are obviously in conflict.
The strategy of the undertaking, biological in nature, is to try to provide an
optimum response to all these conflicting forces in the presence of competitors
who are pursuing a similar strategy but whose responses are not necessarily iden-
tical.  This interplay of conflicts demonstrates both the need for liberty and
that for evolution and creation which, as we have seen in a preceding chapter,
are profoundly anchored in Western culture.  The undertaking presents risks which
seriously commit its members' future since the stake is growth or disappearance.
Participation in this game may be seized upon for the benefit of a few leaders,
but it must not lead to the dilution of authority, in the absence of which there
could no longer be any arbitration between the inevitable antagonisms.

The undertaking must be the place *par excellence* for apprenticeship to social
life.  If its mechanisms were explained to him, the employee would learn to
understand the modern world's complexity; its implications in terms of finiteness
and interdependence; and the need to regard inevitable certain economic and social
constraints.  The undertaking has replaced the parish or the village as the forum
of social communication, a situation which imposes an additional emotional burden
on it.

This description does not solve the problems but it does indicate avenues towards

a solution based on reconciling man with his work.  Comprehension must be facilitated by reducing complexity through decentralization.  Undertakings must be broken up into as many sub-undertakings as are necessary to arrive at production units
or data-processing centres corresponding with the human scale.  This decentralization should be accompanied by far-reaching delegation of powers in such a way that
the largest possible number of employees is associated with the quest for the optimum, which is the undertaking's response to the pressures within and the antagonistic forces which are the expression of its environment.  It is along this line
of thinking that numerous experiments are underway, especially in Germany, The
Netherlands, and Sweden.  It will be necessary to follow up the successes and the
failures while understanding them and promulgating the solutions.

In addition, it is necessary to give back to man in his work a maximum of responsibility in executive tasks while providing favourable conditions for team decisions.
Experiments conducted, *inter alia*, in Italy, inside pilot workshops where only the
overall task is defined, the breakdown of subsidiary tasks being decided upon by
the team actually responsible for doing the job, are a step forward in the quest
for new systems for the sharing of responsibilities which are so necessary for the
cultural genius of Western Europeans.  It is necessary to overcome the contradictions between this search for progress and the maintenance of productivity;  this
calls for both caution and experiment.


### *The Elimination of Distasteful Tasks*

Our citizens have a distaste for certain manual tasks, which they now entrust to
several million immigrant workers who are cut off from their cultural and family
links.  The existence of such a large mass of workers, living as it were in a de-
ported state, is a scandal in the face of the cultural values boasted of by Europeans.  The scandal is all the more striking in that it exists alongside another,
no less painful, namely the presence of an almost equal number of European unemployed.

By means of better organization of labour and more frequent recourse to machines
or automation it would be technically possible today, provided massive investments
in research and development and production equipment were made, to eliminate a large
proportion of the tasks entrusted to what can only be called a "subproletariat".
For the irreducible residue it would probably be desirable either to upgrade various
forms of manual work in such a way that it was accepted by aboriginals, or to entrust to the developing countries which have excess labour on-the-spot execution
of certain manufacturing processes.  The workers would at least experience the immense advantage of finding themselves in their normal family and cultural environment.

This objective of reducing the number of distasteful tasks ought to be considered
as a first priority by the Community.  It would have the additional advantage
that, once the organizational and mechanization effort had been accomplished,
European industry would have made great gains in competitivity.  It would have
progressed from work done by man to work done by machine which, with a few exceptions, is more productive when optimization is properly calculated.


### *Reducing the Fear of Unemployment; Ensuring Virtually Full Employment*

The existence of a large brigade of unemployed is not compatible with the ethics
of the dignity of labour and effort.  Reducing unemployment must therefore be
among those social objectives which enjoy priority.

Studies proposing new solutions ought to be conducted on this subject.  It must
not be believed that a solution is to be found in simply getting the consumption
of material goods under way again.  It is necessary to reconsider the balance of
society, and in particular to study the direct and indirect consequences (not always
as common sense would lead one to imagine) of reducing working hours.  The development of free-time activities could, under certain conditions — in particular as
regards the time and facilities devoted to adult education — help to provide
solutions.

Two mistakes are to be avoided at all costs:

(1)   belief that the peoples of Europe tolerate as readily as the Americans a high
      rate of unemployment;

(2)   failing to increase productivity, despite the technical possibilities of
      doing so, on the pretext that this would aggravate unemployment, especially
      in the services sector.  Europe would become less competitive and so would
      import and consume less and unemployment would increase.


## MASTERY OF COMPLEXITY

One of the principal traits of the post-industrial civilization is complexity.  No
longer is any social, economic, or political phenomenon an isolated one.  A whole
train of secondary effects accompanies it.  Sometimes they are more considerable
than the initial effect.  The fact that the modern world constitutes a finite system — in the sense that no new territories remain to be discovered — adds further to
the complexity.  Decisions, so to speak, bounce back when they encounter the boundaries of the economic and social universe.  The environment is changed by reason
of the very fact that an adjustment has been made to it.

This increasing complexity explains the growth in information activities.  In so
far as this complexity represents progress in co-operation among human beings and
the sign of an improvement in the acquisition and communication of knowledge, we
should rejoice at what we see.  However, when it is a question of growth in para-
sitic activity under Parkinson's law, an effort must be made to combat it because
man's liberty suffers as a result and it represents a tremendous waste of effort.

In this field, too, it might be thought that Europeans are bending backwards to
import from the United States factors making for ossification which the Americans
manage not to amplify, by decentralizing powers and responsibilities, by delegation
of trust, with a smaller workload for the Administration.  In this respect, the
organization of the Commission of the European Communities and its decision-making
channels constitute something which is by way of being a masterpiece of paralysing
complexity.

We must realize that although progress as regards productivity has been considerable
in agriculture and industry, it has been slight in the field of services and the
machinery for collecting information.

In order to improve the quality of life but at the same time reduce collective overheads, it is indispensable to develop two further subjects by study:  reduction of
complexity wherever it is not indispensable and increase in the productivity of
services and of information activities.

The incomprehension born of complexity in its turn gives rise to a feeling of
insecurity.  To compensate, men try to build up systems offering security which,
in the current state of affairs, merely adds to the red tape.  In Europe in general,

and more acutely in the Latin countries, citizens tend to turn to the State and
expect it to behave as an ever-present provider. This desire for security must be
regarded as a major phenomenon of our society. The responses, however, merely tend
to add to the profusion of information, already too voluminous to master, that
dehumanizes relations between administrations and those they administer and the
public services and those who use them. The free citizen is tending to become a
person in receipt of assistance and is thus losing part of his identity; frequently
he becomes a "file" rather than a human being.

Fortunately, as always when a new problem emerges, new tools for tackling it are
being developed. Systems analysis is a new technique that can grasp these phenomena, help to understand them, provide training by means of simulation, and
indicate possible solutions. However, the decision-makers of the current generation have not been trained in the use of such ways of thought and instead of
showing interest and endeavouring to make use of them they are generally distrustful of such intellectual prowess and those who practise it.

### COMPLEXITY AND POLICY

The use of systems analysis to examine this complexity leads us to discard the
two extreme courses of complete economic freedom and centralized planning and
to select a middle road which in fact already exists. There is some reluctance
to use terms such as capitalism and socialism as these expressions mean different
things to different people and are charged with ideological implications.

The very complex system needs to be regulated or, more precisely, to be endowed
with internal self-regulating mechanisms. In these mechanisms, time constants
play the most important role and in an initial analysis, since reality is a
little less simple than its description would indicate, two coexisting systems
of regulation must be distinguished:

(1)  a set of quick-response microregulating devices giving multiple and permanent
     reactions in real time and operating as closely as possible to the initial
     sources of information:

(2)  a set of microregulating devices corresponding to high-inertia general
     phenomena. The microregulating subsystem must have some degree of freedom
     in relation to the macroregulating system. In other words, it must have
     its own constraints and its own ways of optimizing the responses it has to
     supply at any moment to the demands of the environment, without being ex-
     cessively predetermined by the macroregulating system. It must also pass on
     information to the latter and possibly stimulate development.

Adoption of the middle road means that it has been possible to establish fruitful
coexistence of the two regulating systems with neither an excess of planning that
would not give sufficient free rein to microeconomic phenomena, nor complete
liberty for these phenomena where the sum of the individual movements cannot
possibly lead to acceptable macroregulation in a very complex situation.

These general aspects naturally leave very great scope for choices that will
differ according to political options, some preferring a relatively high degree
of planning while others favour greater freedom of competition. The important
point to remember is that it is not possible to achieve anything without giving
plenty of scope to both these forms of regulation. If the Europeans can conduct
studies on this subject, relating them to the scientific field so as to rid the
debate of some of the ideological passion obscuring the issues, great progress
could be made in the social consensus.

# CONCLUSIONS TO PART I

Many personal elements have been introduced into this dynamic study of Western
Europe:  moral and cultural forces moving through history, extrapolations of
observations and trends. descriptions of aspirations or projects.  The reader will
perhaps find it excessively subjective and imbued with a French rather than European
culture;  he should not let this put him off as it is here that we come up against
the difficulties inherent in European diversity.

The opposite reproach could also be made:  this analysis shows a lack of imagination
as it really does no more than extrapolate the curves of the past.  This is true
of the aspirations to economic growth, the extension of the trends towards the
reduction of working hours, the observance of mass production constraints, the
increase in specialized jobs in information processing or communications compared
with agricultural and industrial jobs and more generally respect for the cultural
heritage as it is generally understood.  Although apparently daring, this paper
is fundamentally in line with the forces and trends that have emerged in recent
years.  What is more, the "aspirations and projects" proposed for Europe would
flourish more readily in the United States of America, which is richer, less
vulnerable, and already more committed to some of the courses described, if America
were driven by need.

All this is true, and the reader seeking the seeds of a revolution in the construction of Europe will have to take a different course.  Nevertheless, the extent of the evolution which is proposed must not be underestimated.

A move towards production of durable and repairable goods and the adoption of a
systematic energy conservation policy would involve radical changes in the industrial system that would jeopardize its structures, its methods of winning
customers, and of course the technical specifications for the objects manufactured.

If it were considered realistic to reduce working hours to the equivalent of 32
hours per week (this could be done in a variety of ways, ranging from halving the
working day to the four-day week, sabbatical years, or early retirement), it would
first be necessary to make a prodigious effort to increase productivity in order
to offset the corresponding economic losses and radically to change customs and
habits as the proportion of the non-working population, the unemployed, and those
having no trade, would probably have to be very much lower than today.

If information activities were to become one of the driving forces for growth, Europe would have to make up its leeway on the United States and face up to competition with Japan which is already extensively committed in this special sector.

If free-time activities were regarded as a factor for economic enrichment as well as a factor for the development of human values, there would have to be a greatly developed sense of responsibility as regards the correct use of freedom, new economic and tax incentives, an extensive adult education campaign, a new "cottage" industry, increased pressure for the reduction of time wasted in travelling, and a new housing policy.

If the transfers of technology to developing countries, together with the associated supplies, were to become one of the major spearheads of the Community's export effort, it would be necessary both to establish reciprocal arrangements to provide some of Europe's customers with finance for their imports and to accept that in the long term this would lead to the establishment of fresh competition that would rival us even in the field of advanced technologies.

If consumption habits were to change, with the ambition to succeed by free personal work triumphing over the illusory accumulation of assets, we would move from a civilization of waste and envy, of never-satisfied desires, to a civilization of the self-fulfilled man, responsible and aware of the reward he will receive for his work and his talent, acts and feelings which, in past history, were the privilege of a very small elite.

And if all these projects were to converge, man would be prepared for a civilization which, although relatively frugal in material terms, would have a prodigality of spirit — probably the only practicable solution for a world population numbering 12,000 million amongst whom inequalities would be less scandalously rampant.

These would be the outlines, aims and means of the Second Renaissance.

*  *  *

Why such lyricism now when the first part of this study reached almost discouraging conclusions on Europe's weakness?  Essentially, in order to show that it is not true that the future holds nothing and is bound to be disappointing.  It is not true that we are condemned to zero growth on the way towards underdevelopment. It is not true that there is no longer any social hope; it is not true that the future is predetermined by the present and must lead to dependence, the absence of a political role, and the loss of hope; it is not true that we are moving irreversibly towards the control of complexity by dictatorship.

But it is true that a generation has never been called upon to make such an effort to avert a likely decadence.  The nations of Europe have no chance to achieve this recovery in isolation; they have neither the dimension nor the stamina.  But could not a new impetus be given to the Community with this as its aim?

Pending the implementation of a Grand Project, could we not start with the part that is politically least sensitive but perhaps the most effective in terms of the future, i.e. scientific research and innovation?

Part II
# PROPOSALS FOR MILD TREATMENT

# A COMMUNITY SCIENTIFIC AND TECHNICAL POLICY

<u>INTRODUCTION</u>

In the light of the European balance-sheet, the dangers threatening Europe, and
the ambitions that could be nurtured for a new development, many might be tempted
to press on with a change in structure and policy. The economic war is upon us,
they will say; it has replaced the military war that was the major factor in the
lives of the peoples and their leaders in past history. The battalions of scientists and technicians waging this war for position in the international division
of labour and for control of the salient points in economic power are similar to
the battle forces of the Napoleonic wars. The United States, the Soviet Union,
and Japan have unified their commands; let us mobilize Europe, let us unite, let
us, too, gather our forces and commit them unequivocably to the fight since it is
inevitable. Let us start by concentrating formidable powers at Community level.

We must beware of so radical an analysis. There is every reason to propose more
subtle solutions because the European governments and peoples are far from prepared for a concentration of political forces.

The European is ill-prepared for vigorous treatment. He would like to believe
that his difficulties are benign; he would like to hear that his crises are passing annoyances and balks at recognizing them as symptoms of deep-rooted phenomena
calling for a change of mentality and effort of will. He is therefore prepared
to accept local treatment, administered as far as possible by independent doctors,
but reluctant to accept a vigorous all-embracing approach. Does this attitude
convey sound common sense. a balanced judgement of the real significance of events?
Is it a mere inability to understand the situation as it is? Is it the fear of
changing from a not very effective structure to an even less relevant one? Only
time will show the right answer to these questions. Europe today has need to
mature.

This need to mature does not mean inaction but an effort to prepare the way by
means of mild, well-tolerated treatment exemplary enough to serve as a basis for
wider ambitions when circumstances are ripe. We propose that a start be made
with the Community scientific and technical policy. Why? We see two justifica-
tions:

(1) One is of a practical nature. Research is a long-term affair. Decisions on
    preparations for the future are either electorally favourable[1] or not very
    sensitive. They do not involve irreversible commitments; freedom of choice,
    in the final resort, is preserved. Whatever their difficulties, the Common
    Market partners should therefore find scientific and technical policy one of
    the fields in which co-operation will be easiest provided they do not expect
    the Community organization to take up a position of superiority over national
    policies and attempt to arbitrate, something which no one appears to find ac-
    ceptable.

(2) The other reason was set out in the foreword. The driving force for evolution
    today is scientific research and technological innovation. The agents of
    mutation are no longer genetic accidents but our laboratories and schools
    of thought. Evolution has shifted to the domain of technology and its social
    implications. By guiding and organizing the research apparatus we increase
    our control over evolution; we also give additional scope to freedom of choice
    in so far as — and we must be cautious on this point — man is capable of
    choosing.

---

[1] Result of opinion polls on the receptiveness shown by European public opinion
to science problems (*Eurobaromètre*, December 1977).

It is because of technological innovation that a villager of 1877 would believe
he was seeing a visitor from Mars if he encountered a 1977 camper's caravan with
its transistor radio, television set, cine cameras, washing machine, pocket cal-
culator, frog suit, canned foods, medicines, and airline timetables.  A mutation
of this kind probably establishes irreversible differentiations,[1] which brings up
the problem of the Third World in particularly dramatic terms.  The quality and
intensity of the research and innovation effort will decide Europe's place amongst
the front runners of the "mutants".  That is why it is a fundamental phenomenon;
we shall have an opportunity to revert to these ideas later.

The whole game is being played out in an atmosphere of competition which greatly
resembles the struggle for natural selection.  The European has no choice: to
dispute the reality of this competition is of no practical significance as his
opinion will not influence that of his rivals, but he could perhaps show initiative by selecting his own ground, that of the well-being of man and of social
progress where he is particularly well equipped to win.

## IS EUROPEAN RESEARCH COMPETITIVE?   IS IT CORRECTLY ORIENTED?

Man finds it distasteful to feel that he is dominated by Darwinian-type forces of
biological selection.  He feels that he is and wants to be responsible for his own
evolution.  This is one of the rare values that humanists, Christians, and marxists
have in common; it must be admitted that it is deeply rooted in the heart of man.
To forget this pressure towards the voluntary construction of a better world would
be to commit a grave error in the interpretation of human nature.

It is sufficient to observe the world around us objectively:  selection by competition comes into all compartments of life and more particularly into the rivalries
between commercial firms and between states.  One of the weapons is technological
innovation.  And this race is not about to slow down since, rightly or wrongly,
the tendency to make up for population imbalances by technical superiority is so
strong in some of the developed peoples.

Consequently, the situation in Europe cannot be judged in isolation.  It must be
compared with that of its main competitors — the United States and Japan; for the
time being at least, the USSR only comes into the picture as regards space and
armaments and does not play much of a part in the conquest of capital goods and
consumer markets nor in the international division of labour, although perhaps this
situation is only temporary.

How can the European effort be measured in relation to that of its main rivals
quantitatively, qualitatively, and specifically with reference to the European
identity, its characteristics, and aspirations?

### *Comparison of Quantitative Effort*

Although in scientific research considerations of quality are far more important
than those of quantity, the degree of commitment on the part of political decision-
makers and the economic circles involved can be measured quite significantly by
the expenditure incurred.

The comparison of the budgets of the Europe of the Nine with those of its main

---

[1] These differentiations are not relative to an individual regarded as an educable
person but to man as a centre of communication with his fellows.  It is the group
behaviour and not the individual genetic identity that is mutated, but is this not
just as serious?

market-economy rivals reveals the harsh truth that, relatively speaking, on the basis of the money involved, Europe is in a very poor position.

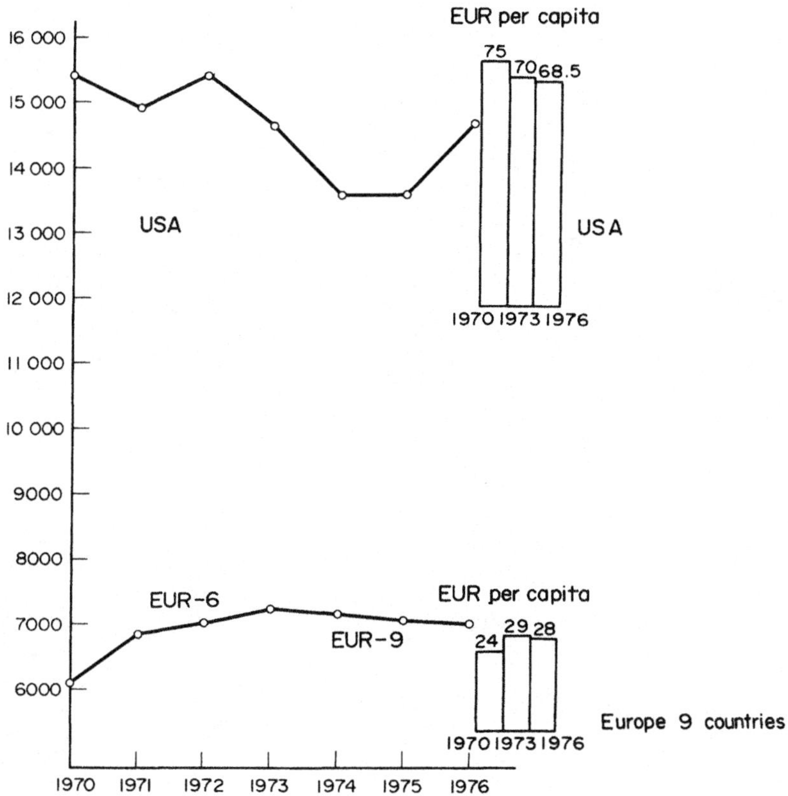


Fig. 27.  Public R&D appropriations USA/EUR-9, 1970-6. 1970 prices and exchange rates.  (Source:  CEC.)

<u>Notes:</u>   (1)   By using the EUR as the monetary unit it has been possible on the graphs to correct the values for the consequences of monetary erosion and consequently to make the investment effort by volume comparable for the United States and Europe of the Nine, despite fluctuations in exchange rates.

(2)   1 EUR = 0.888 6708  gram gold = US$ 1.32 (in 1975).

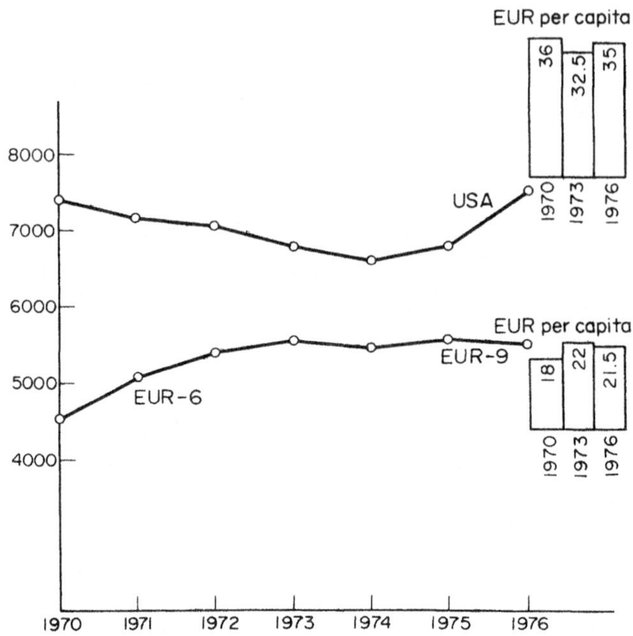


FIG. 28.  Public civil research appropriations, USA/EUR-9, 1970-6. 1970 prices and
exchange rates. (Source: CEC.) <u>Notes</u> (as for Fig. 27.)

<u>Table 8   Comparison of budget estimates for 1977</u>

| United States | Million dollars |
|---|---|
| Government-sponsored research | 23 500 |
| (of which military research) | 12 000 |
| Industrial and other research estimated at | 20 500 |
| Total | 44 000 |

Equivalent to 2.3% of the GDP


<u>Europe of the Nine</u>

| | | |
|---|---|---|
| Public expenditure: | 10 400 million EUR or approximately | 13 000 |
| Expenditure by companies and non-profit-making associations: estimated at 10,000 million EUR or approximately | | 12 500 |
| | Total | 25 500 |

Equivalent to 1.6% of the GDP

Between Western Europe and the United States the figures shown in Table 8 reflect
the volume of expenditure fairly accurately as the costs per research scientist
employed in Europe are today very close to the American level.  As the United
States has a smaller population than Europe of the Nine, the comparison of per
capita expenditure is even less favourable, as Figs. 27 and 28 show.  As a rough
approximation, the volume of Research and Development investment per European
is less than half the American level.

If budget trends in recent years and estimates for 1978 are examined, the situation
is found to be worsening; the United States has been stepping up its research
effort since 1976 while the Nine have been tending to cut theirs or at best to
keep it constant, because of the priority given to the fight against inflation
(see Figs. 29 and 30 for the American estimates).

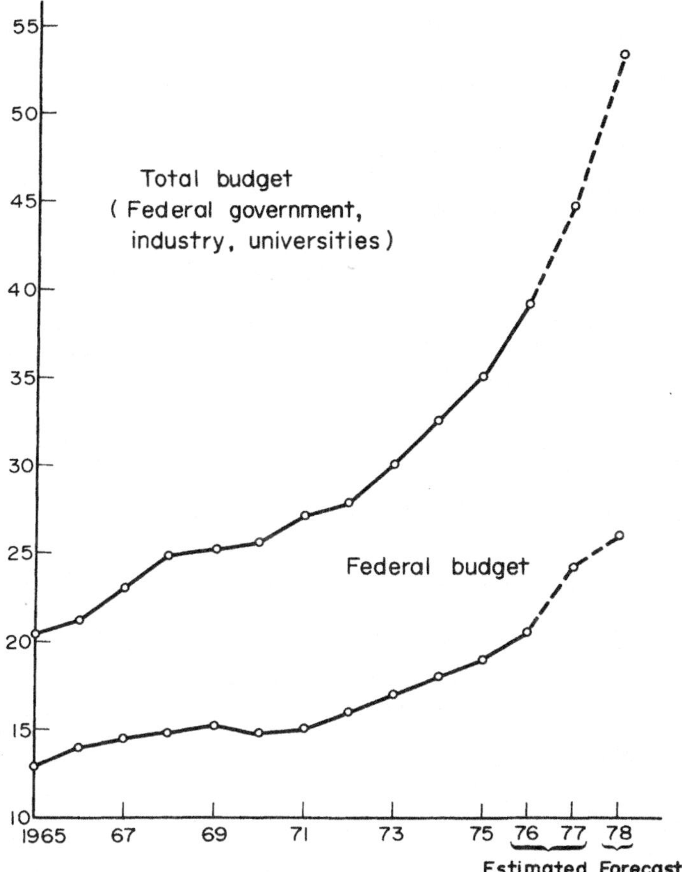


Fig. 29.   United States R&D budget (current dollars) since 1965.
           Equipment and operations, excluding construction.  (Source:
           NSF, Battelle Institute.)

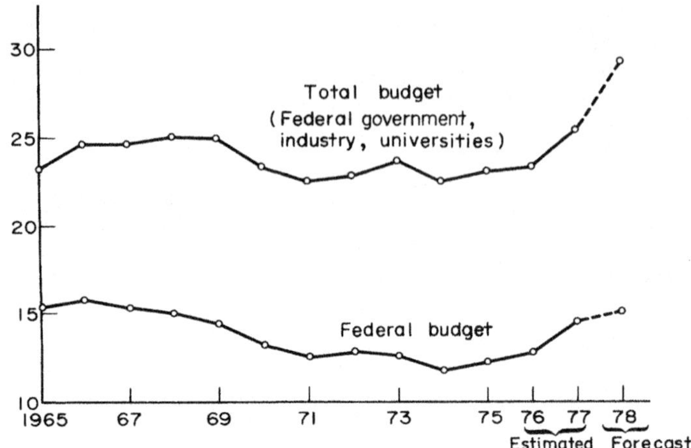



Fig. 30.   United States R&D budget (constant dollars, 1968 = 1) since
1975.   Equipment and operations, excluding construction.
(Source: NSF, Battelle Institute.)

A valid comparison with Japan cannot be made on the basis of budget appropriations[1]
as for various reasons the apparent unit cost of a Japanese research scientist
is very much lower than the cost of a European or American research scientist.   To
have an idea of the effort expended, the number of persons employed provides the
best basis for comparison (Table 9).

Table 9. Comparison of the number of specialists working on R&D

|  | Japan (year 1975) | Europe of the Nine |
|---|---|---|
| Research scientists | 255 000 | 260 000 to 290 000 |
| Research assistants | 84 000 |  |
| Technicians | 88 000 | approximately 300 000 |
| Other | 64 000 |  |
| Total | 491 000 | approximately 570 000 |
| Population | 111 000 000 | 258 500 000 |
| Number of research scientists per capita | 0.23% | 0.11% |

Sources:   Japan — French Embassy in Tokyo.   Europe of the Nine — Commission
of the EC.

---

[1] It is, however, important to know the total sum allocated in 1975 to research
and development (DIBRD).   This was 2 979 000 million yens, or approximately
$10 000 million.

Official documents reveal two typical features of the Japanese effort.  One is the very small, not to say non-existent, share of innovation for military purposes, and the other the speed of growth.  In the five years 1971–1975, 61 000 new scientists started work in Japan, a number approximately equivalent to the total research staff employed in each of the major Community countries.  Japan's priorities are fairly clearly revealed by the major items in its budget appropriations: nuclear science, information technologies (space, telecommunications, informatics; environment, oceanology.  In the information technologies a large part is played by studies of automation and robotics to improve the productivity of industries and services.

*Can the Quality of European Research be Evaluated?*

A qualitative comparison is more difficult and, of course, more open to challenge. Consequently the opinions given should not be too dogmatic.  However, the general agreement accorded to President Sune Bergstrom's warning justifies the following long quotation.  This speech was made on 10 December 1976 at the traditional Nobel prize-giving ceremony in Stockholm.[1]

"The distribution of prizes amongst the various countries is an interesting reflection of the changes that have occurred at international level in the work on a research subject.  A review is even more interesting this year in that all the prizes including the economics prize in memory of Alfred Nobel went to citizens of the United States.

"The world press has commented widely on this fact in terms going well beyond ordinary information for the general public.  For example, some authors have concluded that the Nobel Foundation wished to contribute to the Bicentenary celebrations in the United States, while others saw it as a political move by the new Swedish Government. . . .

"How do the changes in the distribution of the prizes reflect changes in research activities throughout the world since 1901?  In the early decades of the century, the large majority of the prize winners were Europeans.  Only four of the eighty-four prizes awarded in the first twenty-five years went to Americans. The number of American winners began to increase in the thirties, and no less than forty-eight of the hundred prizes awarded in the last twenty-five years have gone to them.

"Since the war the United States has received at least one prize every year, with two exceptions.  Four times they have taken all three prizes for physics, chemistry, and medicine.  In terms of pure probability, it is therefore hardly surprising that the literature prize — already awarded eight times to American writers in the past fifty years — and the economics prize in memory of Nobel should also go to the United States at the same time as the scientific prizes, as was the case this year.

"There are, of course, many reasons for the great predominance of American research since the thirties, quite apart from the ravages of two world wars and their numerous political consequences that have greatly hampered developments in Europe.

"In the mid-thirties, the American Government began to be aware of the importance of research for economic development, and as a result Congress voted a substantial increase in the funds earmarked for research activities by the

---

[1] In 1976 all the Nobel prizes went to Americans: chemistry to W.N. Lipcomb; physics to B. Richter and S. Ting; medicine to B.S. Blumberg and D.C. Gajdusek; economics to M. Friedman; and literature to S. Bellow.

Federal authorities.

"The United States opened up its lead mainly during the Second World War with
the successful results of research and development work backed by military organizations, culminating in the atomic bomb, radar, antibiotics, etc.  In the
twenty years after the war, the experience gained led to an unprecedented development of civil research in the United States by the National Science Foundation and, in the biomedical field, by the National Health Institute.

"For a quarter of a century these efforts have put the United States in the
forefront of research in many fields.  Quite apart from these economic conditions, however, other factors have played a great part in America's rapid growth
in the research sector and are perhaps of special interest to Europeans.  Universities developed rapidly in the fifties and sixties all over the world.  In
many countries, starting with Europe, this development took place within the
traditional rigid hierarchical structures.

"In the United States, on the other hand, the growth of university research
took place in a dynamic and outward-looking atmosphere and in a form that
might be termed democratic as far as research workers were concerned.  Many of
those who visited American institutes and attended scientific congresses after
the war were struck by the natural manner in which teaching staff and students
held scientific discussions on an equal footing and were surprised by the
practice of entrusting young scientists at the beginning of their careers with
responsibility for independent research activities within large bodies.

"The second important factor for the United States was quite simply the size of
the country and the resulting high mobility — it is easy for a young scientist
to find a research group or context that suits him, particularly at a time when
the mobility that had earlier existed between European countries had almost
totally disappeared.

"All these factors, combined with liberal immigration laws, made the United
States a refuge for numerous scientists who left their own countries for political reasons or because of lack of support or of favourable working conditions.

"A large number of European scientists therefore emigrated to the United States
especially in the thirties and forties.  Some of them, and more recently some
of their pupils, are to be found amongst the U.S. prize winners.

"In the past ten years, however, there has been considerable progress in Europe
— as elsewhere — in the understanding and appreciation of science.  The
results are already evident, although much remains to be done.

"The American predominance has most probably reached its peak and it is to be
expected that in the coming decades the other industrialized countries will
increase their research and development potential.  A pointer in this direction
is the fact that one of this year's physics prize winners did much of his work
at CERN, the European Nuclear Research Centre in Geneva."

Should Europeans see the moderation of his words and the relative optimism of his
conclusion rather as a desire to be courteous than as holding out real hope for
their future?

If, leaving basic research, we now turn to industrial research and its links with
university research, the findings are not much more reassuring, as we have already
seen.

Judged on the criteria of mobility, links with the economic sector, and communication structures, the conditions under which European researchers work are on average inferior to those enjoyed by their American colleagues. As regards dissemination of innovation and its profitable exploitation, the U.S. market, which is unified and remarkably well protected among the advanced technologies, constitutes a sort of echo chamber or amplification system which is missing in the divided and compartmentalized European market. The latter seems incapable of organizing temporary protection to foster the endogenous development of certain sensitive products. These handicaps are aggravated by the development of multinational companies. As already mentioned, these companies are excellent instruments for the exploitation and dissemination of innovation, but whereas they were almost exclusively European up to 1950, now American or Japanese companies tend to predominate.

Japan has up to now been remarkably successful in combining its innovation effort with the implementation of an industrial strategy designed to concentrate its endeavours on climbing to the top of the tree in several key sectors in turn. On the other hand, the European Community has been incapable of adopting an industrial policy.

As against this, more optimistic arguments can be put forward without too much fear of contradiction. The intellectual seedbed of Europe, nurtured over several generations, is still remarkably fertile; the level of knowledge is extremely high. Whenever ambitious targets have been set and adequate material resources provided, the projects have proved successful, at least up to prototype or preproduction stage (nuclear science, aviation, chemistry, pharmacy, etc.).

These successes prove that there is a resource available. They show that all is still possible and that Europe is capable of recovery. However, if investment remains at half that of its main rivals and if quality is not improved, this relative wealth will drop to the point where Europe can no longer remain competitive. This serious warning is given without reservation.

*Are the Aims of European Innovation Policy in Line with the Characteristics of the European Identity and its Aspirations?*

It is difficult to reply to this question because there is no declared European policy on scientific and technical research. The Commission of the European Communities handles research appropriations equivalent to less than 0.6% of the total expenditure; on this basis it obviously cannot conduct a policy. Most of the Member States have not published guidelines for their innovation effort. This lack of theoretical planning does not necessarily mean that there are no practical guidelines.

It appears consistent with market economy principles for a substantial part of the innovation effort to be carried out by industrial and commercial companies. Research and development planning is carried out at company level; this means that it is a part of the strategy of commercial operators covered by trade secrecy. Rivals endeavour to use the weapon of technological progress to surprise their competitors by launching new products and introducing more productive methods.

It is obvious that these strategic factors step up the pressure towards innovation in order to solve problems of competitiveness in the short and medium term. It is not at all obvious that the combination of these activities could cause society to evolve towards the fulfilment of projects involving better adaptation to the conditions of the social or economic environment that can be foreseen for the

long term.  That is why governments participate in the funding of research and
development; they wish to attain certain objectives regarded as being in the
national interest.

During the period 1938-1945 the war effort took up most of the resources assigned
to applied research and the development of prototypes and preproduction models
in pursuit of the aims of national defence and attainment of a self-sufficient
economy.  With the coming of peace it appeared that the initiative in the way of
technical innovation would once again fall to private investors.  In fact this has
only partly been the case; in the Western industrialized countries the government's
share amounts to approximately 50% of total expenditure, both in the United States
and in most European countries.  This 50/50 estimate, which is no more than ap-
proximate[1]  shows that a State scientific and technical policy accompanies, and in
some cases replaces, the strategy of the private companies.  Funds from government
budgets go without distinction to public and private laboratories.  Going to the
extreme, one could imagine research budgeting deriving primarily from public funds
while the work was carried out for the most part by firms or private non-profit-
making associations.  Europe differs from the United States and Japan in that a
larger proportion of its research work is carried out in research centres that are
either state owned or under the direct responsibility of the State.  Paradoxically,
this situation reduces both the power of the Government to select objectives and
the efficiency of the transfer of the knowledge acquired in the laboratory to those
responsible for economic and social utilization of the results.  We shall revert
to this paradox later.

In this interaction of private enterprise and state policies, what can be said in
respect of quality as regards the direction of the European innovation effort?

A look at the priorities adopted for the various programmes reveals the predominant
influence of two main factors:  sensitiveness to the scientific and technical mode
that has come from the United States and, especially since 1973, the concern with
survival.

It is difficult to affirm a European identity in the face of the American influence,
radiating an extraordinary success, because Europe, culturally very close to its
American  daughter, is caught up in a technological and social mutation of a similar
nature.  The European Community could take on the task of injecting specific information as a type of hormone to protect itself against identification with the
United States whose problems, as we have seen, are different at almost all the
practical levels: politics, geography, natural resources, economics.  The aim would
not be to engage in a struggle to affirm an identity in a manner antagonistic to
the United States but to throw light on and promote points that are complementary.
Insufficient efforts are being made on these lines today; they appear only in
isolated areas of national policies or in the course of some international
co-operative projects.

Since 1973 a policy of "survival above all" has been developing in Europe, but
the serious oil threat has not resulted in any combined mobilization of material

---

[1] Innovation depends on basic or fundamental research; it then goes through the
stages of applied research and applications research, followed by development,
after which the mock-up, prototype, or preproduction stage is reached.  The process continues with demonstration and sales promotion.  It is pointless to try
to demarcate strictly these various stages; they overlap and intersect, forming a
continuous process with frequent feedbacks.  This is why it is so difficult to
agree on how to express the cost of research or even the number of research workers.
However, comparisons are valid in terms of orders of magnitude.

and human resources that would have impressed the rest of the world.

The effort to survive is essential, but this defensive attitude is not sufficient;
it creates no enthusiasm, and experience shows that it does not rally forces.
Poorly understood by public opinion, it does not prepare the way for the sacrifices that should be made today for the benefit of tomorrow.  It does not encourage financial investors hostile to anything lacking in ambition and future prospects.  That is why we consider it so necessary to add the idea of an offensive
movement aimed at surviving to live a better life.  We shall see how this idea
can be taken up in the selection of priorities, but we must first try to understand why the European decision-makers appear to have little liking for scientific
and technical effort.  Are there reasons for challenging science?


<u>SCIENCE CHALLENGED</u>


*The Facts*

The need to make savings so as to support currencies, improve trade balances, and
reduce State spending does not fully account for the fact that budget appropriations for scientific and technical research in Western Europe have remained sta-
tionary or even declined since the beginning of the 1960s.  Other factors that
deserve some consideration have helped to strengthen this trend.  It is difficult
to understand why Europe as it is today, with all the threats hovering over it,
should take the risk of being outstripped by its rivals in the only specialized
field where it is not seriously handicapped — brainwork and inventive capacity.

It cannot be denied that science and technology no longer enjoy the untroubled
atmosphere of the immediate post-war years.

Some scientists are concerned at the use made of their discoveries, in particular
as regards the destructive power of weapons.

Public opinion is alarmed by the environmentalist movement, the sentimental force
of which should not be underestimated.  However, the ecologists' messages are not
always properly understood; their warnings that priority should be given to safety
are too frequently seen as a rejection of technological development which is identified as the cause of hazards.

After the scientific craze of the 1960s, governments have been disappointed at the
low return they have received both from the scientists themselves and from their
results in the way of solving economic crises and the problems of society.

What is more, Europe is once again imitating the United States while running
several years behind; this lack of initiative will eventually ruin it.  The U.S.
Government cut back its research and development effort in about 1970.  Instead
of catching it on the wrong foot and moving into the attack, most of the Community
Member States have followed the fashion set across the Atlantic.  However, this
fashion never caught on in American industry, and two years ago its direction
changed; have the Europeans noticed this?


*Questions*

This situation gives rise to a number of questions:

(1)  How can we learn to attribute due importance to long-term preparations?
     Market economies and democracies are particularly vulnerable to the tempta-
     tion to live for the moment.  Heads of companies have to pay great attention
     to their end-of-year balance-sheet, while politicians have to submit them-
     selves frequently to the whims of the electorate.  Would not one way of over-
     coming this obstacle be to base scientific and technical policy on organisa-
     tional structures attached to the highest level of government and business
     and to make multi-annual budgets a regular practice?  At the highest levels
     there is a particularly keen sense of responsibility for the future; decisions
     in favour of long-term investment are the most readily given.  If, in addition,
     the public were kept adequately informed, scientific and technical policy
     would be expressed in terms of electoral concern.  Is this wishful thinking?

(2)  How is it possible to prevent the misgivings of research scientists being
     confused with the unrest in the universities?  The latter difficulties stem
     from the fact that the universities are ill-suited for mass education; the
     phenomenon is largely independent of research.

     Researchers will always be sensitive.  They are by nature inclined to judge
     the present by what they expect of it for the future; consequently their
     opinions are not a matter of common sense.  They must at all times fight to
     protect their originality; their ideas on liberty are particularly intransi-
     gent, as is clear in countries where there is not complete freedom of move-
     ments or expression.

     The misgivings of research workers are also expressed in defensive reactions.
     If a serious opinion poll is carried out amongst circles close to — but outside -
     research, the corporative egoism of researchers is criticized.  "Career se-
     curity", it is said, "is of greater interest to them than their duty to
     society".  "They work to satisfy their own curiosity rather than to serve
     others whose future depends on their work."

     This might appear a mere extravagance of language that is difficult to un-
     derstand and still more to accept, but unfortunately it does express the true
     feelings of those who are responsible for formulating political and economic
     decisions.[1]  When a research centre has gone through a crisis and when the
     crisis has been overcome by creating public awareness — the history of the Ispra
     Joint Centre provides an example — it is extremely difficult to erase the memory
     of the difficulties from the minds of observers.

     Is it not up to researchers themselves to come to grips with this situation
     and examine its foundations?  Important movements are afoot to study the
     responsibility of scientists to society; should they not be encouraged by
     demanding that all aspects of responsibility be taken into account, including
     the immovable constraints arising from the economic environment?

(3)  What should be the response to the attacks of those who believe technology
     to be dangerously extravagant or who regard technological development as
     contrary to the interests of man?  Two arguments are readily developed; they
     are heard quite frequently in intellectual circles and amongst members of
     the liberal professions.  They encourage public opinion to reach disparaging
     judgements on scientific research; they provide decision-makers with justi-
     fication for cutting budget appropriations for research and development.

----

[1] Opinions of this kind are expressed both in the West and in the East.

Some consider that technology is coming to the end of the easy production
stage; the richest veins of science have been exhausted, and to exploit a
lower grade of ore the sums to be invested are too large in relation to the
expected yield.

Others consider that technological development has been too fast; it is no
longer progress, but threatens to destroy the equilibria of the living world.
It does not improve man because it proposes ways of life that cut him off from
his cultural roots.

Should not scientists and technologists encourage discussion on these matters?
They are serious; in some respects they raise real problems which are not
unconnected with the concerns expressed in the first two parts of this study.

To respond with adequate information it would be necessary to bring the current situation of science and technology to the knowledge of all in a suitably
drafted form without using the jargon that makes communication so difficult.
This is not the purpose of this study, and consequently I merely refer the
reader to the following list, obviously not exhaustive, of the scientific
specialities that at the present time hold out great hopes for the future.

The following are fields in which major technical breakthroughs are to be expected
in the next thirty years (with no claim to be exhaustive):

<u>Energy</u>

New sources: solar, geothermal, wind, waves, etc.
Hydrogen (transport, storage).
Nuclear (fast neutrons, breeder reactors, fusion).
Coal: gasification, synthetic oils, new mining and transport techniques, etc.
Bituminous shale.
Bio-energy.
Energy conservation: insulation, recycling, etc.
Use of lasers as local sources of energy, fusion, transport, regulation, etc.

<u>Oceans</u>

Mining of the ocean bed (mineral nodules, etc.)
Mining of the continental shelves.
Aquaculture and photosynthesis in a marine environment.
Desalination of sea water.

<u>Space</u>

Applications for telecommunications, broadcasting, television, and data
transmission.
Applications for remote sensing.
Meteorology.
Navigation aids.
Scientific experiments in space.
Other (solar energy, production of extremely pure products, etc.)

<u>Preservation of the ecology, conservation of natural resources</u>

New recycling techniques applied to industrial and household waste, precious
metals, etc.
Biodegradation and bioregeneration techniques.

Antipollution devices at source (vehicles, factories, etc.).
Systems analysis applied to the cycles of scarce raw materials, pollution study, etc.

## Understanding of major natural phenomena

Seismology, vulcanology, meteorology.
Theory of the earth's tides, tectonic structures, etc.
Climatology, study of solar tides.
Study of the phenomena that protect the upper atmosphere and the ionosphere.
Theory of atmosphere/ocean exchanges, global energy balances, etc.
Monsoon analysis and prediction.

## Life sciences – biology, medicine, agronomy

Development of recent acquisitions of biology (molecular biology, genetics, enzymology, etc.).
Virology, immunology, etc.
Animal and plant hybridism.
Tropical agriculture.
Emergence of the bio-industries (pharmacology, nutrition, biological synthesis products, etc.).
Evolution of medicine under the influence of micro-engineering, model building, computers, electronics, etc., and as a field for the application of biological advances.

## Information technologies

Application of new materials and components: glass fibre, lasers, microwave and opto-electronic components, high-speed macro-memories, highly integrated circuits, holograms, etc.
Microprocessors; computer networks.
Artificial intelligence, robotics.
Automation of work in the services sector and office work.
Teleconferencing, electronic mail, facsimile reproduction, teletypesetting, videophones, etc.
Audiovisual techniques (video-discs, tapes, cartridges, etc.)

## Transport

Supersonic and/or low noise aircraft.
New levitation devices (air cushions, magnetic fields, etc.).
Systems analysis applied to public transport, traffic control, etc.
Application of electronics and computers to the motor-vehicle.
New safety techniques.
Multimode urban transport systems.

## Organization – planning – social life – communications

Scientific study of social organizations.
Applications of systems analysis to the study of regional planning and housing (humanization of the urban environment, optimization of the cost of sites, construction and maintenance, etc.).
Creation of new economic indicators expressing the degree to which human requirements are satisfied, the quality of life, etc.
Study of economic phenomena (inflation, currencies, etc.).
Study of procedures to combat unemployment.
Modelling of resources problems (population, nutrition, energy, raw materials,

etc.) and development of forward studies based on scientific methods and
computer facilities.
Progress in sociology, psychosociology, political science.
Use of tools (mathematics, computers, etc.), to aid decision-making.
Studies on new methods of education, teaching, learning, etc.
Application of statistical analysis and computers to human sciences, historical
studies, linguistics, law, etc.
Tools for the machine translation of languages.
New man-machine communication methods (speech recognition and synthesis,
automatic document reading, etc.).

* * *

A point-by-point examination of these special subjects would lead to the following conclusions, which we ask the reader to take on trust:

(1)  It is incorrect to say that science has exhausted its most fruitful resources.
     On the contrary, we are experiencing the emergence of the "life sciences"[1]
     which are concerned with the action of micro-energy phenomena.  These life
     and information sciences can be studied by small teams, and yet most probably
     herald new and far-reaching developments in a wide variety of applications at
     low cost.

(2)  Another very fruitful area is multidisciplinary research which brings together
     several special subjects.  Examples are the combinations of electronics, data
     processing, applied mathematics, and health techniques; of biology and chemis-
     try; of mathematics, systems analysis, data processing, and the human sciences,
     etc.

(3)  At the present stage in human development, with the 8 to 12 000 million people
     that will populate the earth in the next century, only technical progress
     can counter the genuine threats of imbalances.  The solution cannot be to
     stop short; this would be equivalent to standing in mid-stream in the most
     vulnerable position.  Nor can we retrace our steps, however rosy (and this is
     open to discussion) the past may appear.  The only course that offers any
     hope of providing a decent life for such a large population is to make rapid
     progress in numerous fields: nutrition, new sources of energy, better pro-
     tection of the environment, conservation and recycling of scarce resources.
     A new surge forward in scientific knowledge and correct application of tech-
     nical innovation are the only ways of attaining this.

     To bring this home to public opinion, should not scientists devote much of
     their time to information, education, communication of their knowledge in
     terms that can be understood by all?  A technological breathing space, which
     would deliberately delay practical utilization of scientific results, would
     run counter to the humanitarian intentions of those who advocate it, as it
     would aggravate the crisis caused by the population explosion and the
     deterioration of the environment.  How can this be put across?

(4)  What reply should be given to those who advocate marking time so as to leave
     the financial burden of innovation to our economic rivals, later selecting

---

[1]  The term life sciences will be used on several occasions in this study.  It
applies to those sciences which study living systems in both their biological and
their societal forms (societies of insects, human societies, individual and
social patterns of behaviour) using systems analysis and modelling.

from their results those which would be most useful to us?  This apparently
intelligent economic calculation conceals a tragic error.  To reveal this
error, researchers, innovators, and manufacturers have no arguments to fall
back on other than their own experience, which in part is impossible to put
across.  One must have had firsthand experience of the ups and downs of the
competitive struggle for innovation to realize that it is very difficult to
jump on a moving train.  This is not merely intuition acquired on the job.
It is a fact that can be expressed in specialized technical language, case
by case, and above all in econometric terms.  However, this specific technical
language inhibits communication with those who have to be convinced, the man
in the street who reflects public opinion, the mass media which form that
opinion, and political decision-makers who are distrustful of specialist
vocabulary.  Perhaps examples will be sufficient to make the point.  Past
experience with Japanese electronic watches, steel, and shipbuilding, with
American computers and electronic components demonstrates more convincingly
than any talk that a place on the rungs of the ladder of the international
division of labour are hard won and that it is very difficult to dislodge
those who gain the first foothold in markets which they have won by their own
pioneering efforts.

In brief, there is no justification for rejecting the scientific and technical
research effort.

(1)    The source of technological development has not dried up and drawing on it
       is no less rewarding now than in the past.[1] On the contrary, the prospects
       offered by the emergence of the life and information sciences and of multi-
       disciplinary research are particularly well-suited to the European bent.

(2)    Everything militates in favour of a policy reflecting initiative and pio-
       neering spirit.  By taking the offensive, we avoid being left behind, some-
       times with no hope of catching up; we oblige others to follow us and we pre-
       serve our liberty.

(3)    There is no future in deliberately bringing scientific progress to a stand-
       still and blocking innovation.

These conclusions are not derived from theory but from an analysis of the facts.
The author would have been very happy to propose a method that is less costly,
in terms of the sacrifices to be made, to prepare for the long term at the expense
of the short term; he would have liked to invent a shortcut avoiding the need to
run the full distance.  Unfortunately, facts are more compelling than desired.  It
is necessary to learn once again to relish research, to associate it with a project,
and to select priorities.

*The Gap Between the Assessment of Intellectuals and Public Opinion*

An opinion poll carried out under the auspices of the Commission of the European
Communities, following a proposal by Professor Prigogine,[2] has shown that
European public opinion is more or less unanimously in favour of scientific
research.  Faith in the virtues of science as an instrument for human progress is

---

[1] It must be acknowledged, however, that with the increasingly cumbersome administrative structure of research and the high level of international competition it
is becoming increasingly difficult to acquire superiority in the order of knowledge.

[2] 1977 Nobel prize for chemistry — member of CERD.

deeply rooted in the minds of the people, whatever the age, background, or nationality of the Europeans concerned. At the same time, however, a majority opinion is expressed in respect of dangers created by technological innovation. This desire for research to continue, heavily subsidized by the states and receiving substantial aid from the Community, contrasts strangely with the scepticism expressed in some intellectual circles and the lack of interest shown by political decision-makers. Who would dare to say in this instance that public opinion is wrong? In that case, why regard it as of such little significance?

## AN ATTEMPT TO DEVISE A STRATEGY OF PRIORITIES

As regards research objectives, the desire of any political authority, probably shared by public opinion, is to determine a limited number of priority fields. The task would be easier if it were possible to eliminate once and for all exhausting and constantly criticized choices between medical research, progress in agriculture, in data processing, telecommunications and electronics, new sources of energy, and a hundred other varied subjects, all supported by pressure groups. Unfortunately, the development of the technical society is accompanied by a widespread growth in very varied fields, none of which can be ignored. To accept this complexity is the first useful step in thinking about science and European innovation.

The European Community is a vast consumer market, second only to the United States in quality and quantity. There is an immense diversity of traditions, trade patterns, and requirements. It would, therefore, be wrong to make choices by exclusion. The Europe of the Nine cannot simplify by substraction; it must conduct scientific research in all specialized fields, at least from a defensive and vigilant position if not from an offensive position of creative originality.

A scientific and technical research policy cannot ignore the diversity; it will, therefore, have to support a minimum effort in all specialized fields. In particular, it will, at national or community level, have to devote sufficient funds to basic research, which is essential for good-quality higher education and is the source of all innovation.

It is, however, necessary to go still further. Without being diverted by the storm of protest that any proposal of this kind is bound to provoke, we could try out the following procedure for selecting priorities.

(1) Some subjects hold out rich promise of innovation potential because recent discoveries in the area of fundamental knowledge have opened up further possibilities of applications in the near future. They may be said to offer virgin land for man to conquer. In 1978 this applies to everything concerning the life sciences and information technologies.

(2) Some subjects are very critical to our economic survival.

(3) Another set of priorities is relevant to the aim of ensuring a better life.

Before clarifying the content of these three categories of objectives by going into greater detail, it may be advisable to take an overall view. This is summarized in Table 10.

Table 10  Contents and criteria of the priority fields

| Category of objective | Content in order of importance | Major criteria justifying priority |
|---|---|---|
| **I**<br>To profit from the upswing in the life and information sciences | 1. Biology<br>2. Applications of mathematics, systems analysis, statistical analysis, and data processing to futurology and decision—making<br>3. Social sciences | Resources required are moderate<br>Guarantee of success provided by the enormous European cultural heritage<br>Relevant scientific research on the verge of decisive progress in methods or results<br>Specialized subjects likely to promote new industries with a low consumption of scarce materials |
| **II**<br>To organize survival | 1. Energy<br>2. Improvement in service productivity<br>3. Competitiveness of agriculture and industry | Reduction of economic dependence<br>Reduction of costs and mastery of complexity<br>Innovations specific to the European identity (shortage of space, vulnerability of the environment, shortage of raw materials) |
| **III**<br>To take the offensive for a better life | Establishment of reciprocal economic links with the developing countries | Stimulation of Europe's economic growth by the development trends in countries that are today economically poor but are rich in real requirements |
| | 1. Information technologies<br>2. Development of economic freetime activities | Meeting specific needs of highly industrialized countries<br>Activities in which the value added by brain power is high and the consumption of scarce materials low<br>The emergence of new consumption models |

In Table 10 participation in the progress of the developing countries comes under
both the "survival" and "better life" headings.  There is probably no need to
justify this.  In actual fact, the three categories are inseparable.  The life sciences will supply the keys for the new innovating industries and for the improvement of social life, whereas "survival" is obviously an essential condition for a
"better life".  The author believes, however, that the fight for survival will
never be truly won unless the European peoples can be given the hope of a better
life.

*Means of Profiting from the Upswing in the Life Sciences*

Reference has already been made to the emergence of the life sciences, which are
concerned in particular with the nature and behaviour of animate beings.  At
elementary level, these sciences are concerned with phenomena controlled by low

energy changes, unlike the natural sciences, of which physics and chemistry are the
major branches.  This concept of energy is extremely useful in understanding their
special nature.  When important mechanisms are triggered by very low energy transi-
tions, as is the case, for example, in genetic memory or the functioning of the
nerve cell, the dominant phenomena are no longer governed by global behaviour but
by fluctuations[1] This accounts for the fact that the life sciences cannot be reduced
to simplifying apparently immutable laws like the mechanics of celestial bodies or
the physical chemistry of minerals, but are sciences of change, complexity, and
interreactions; the laws of thermodynamics are, of course, observed but do not in
any way help to predict individual modes of behaviour which are, by their very essence, uncertain.  What is involved here is indeed a combination of change and
necessity; we shall have an opportunity to revert to the conclusions that can be
drawn from this concerning the structure and mode of action of scientific research.

Biology is the first expression of these sciences of complex energy microphenomena
but, by extension, it is not artificial to include among them the human sciences
that also embody systems characterized by complexity of interreactions and by change.
For long a tool of the physicist, mathematics is starting to tackle the description
and modelling of these phenomena, which involves it in a radical transformation.
This leads to a development of tools and concepts that hold out hope of very great
progress in fields that up to now have been restricted to observation.

There is every indication that recent biological discoveries hold out hope of
applications of enormous scope.  Prospects are comparable to those offered by
physics at the beginning of the century.  The breakthrough must be followed up, and
it would be all the more unpardonable for Europe to show no ambition to succeed here
since the financial resources called for are relatively modest compared to those
required for other fields of research such as high-energy physics, the economic
and human impact of which are less obvious.

Incidentally, it is extremely fortunate that these life sciences — or sciences of
change and complexity — are holding out such excellent prospects at the very time when
the post-industrial society is becoming bogged down in an excess of complexity and
a failure to understand change.

Generally speaking, the advent of the life sciences will once again make small,
highly motivated teams efficient; the large laboratory organized more or less like
a factory is now valid only in exceptional circumstances.  Multidisciplinary organization will prove most fertile.  If knowledge is to progress, calls must be made
on many disciplines:  mathematics, model-making, scientific calculation, electronic
and optical instrumentation, chemical analysis. psychology, sociology, etc.  The
best results will, therefore, be obtained by ad hoc operational groups, task forces
set up to organize "Laboratories without walls" cutting across organic research
units so as temporarily to bring together experts working in different disciplines.
This meeting between basic and applied researchers of various origins, who will
have to learn each others' languages and pool their knowledge, will not come about
spontaneously.  It must be catalysed.  If this catalysis is done judiciously, enthusiasm, discovery, and invention will spring forth from stocks of knowledge that
are barren in isolation.  We shall come back to the structures required for the catalysis of scientific and technical research.

---

[1] Fluctuation means the sudden appearance of an unlikely phenomenon constituting a
significant departure from the normal.  This anomaly can have appreciable secondary
effects, e.g. cause genetic mutation or disturb a delicate structural balance.  It
is not generally possible to predict a significant fluctuation, to evaluate its
probability, or to imagine its consequences.

Is there any need to say that Europe, with its high population density, the communication facilities that geography and history have provided for its peoples, and
the wealth of its cultural heritage, appears remarkably well placed to succeed in
these disciplines?  There are still many obstacles to be overcome, however.  These
include the individualistic nature of the European research scientist and his reluctance to accept the constraints of the "laboratories without walls"; the gulf in
Europe between scientific discovery and utilization of the results by industry;
misunderstandings regarding the concept of multidisciplinary research; and the
remoteness of research in social sciences from the real problems of life in industry
and the public service and the tendency of such research to resort to intellectual
concepts that are difficult to put across to non-specialists and are perhaps
pointless.

Giving priority to these special subjects means providing the means to overcome
all these obstacles, which will not be possible without allocating extensive human
and financial resources.

*Organization of Economic Survival*

(1)  <u>New energy sources and energy conservation</u>

Whatever standpoint one adopts, energy will be in the forefront.  The main factors
justifying this priority include:

(a)  Dependence resulting from the cost of oil imports.

(b)  Growing scarcity (20 to 50 years) of liquid and gaseous fuel resources with
a low cost of extraction, and the vulnerability of their political environment.

(c)  The Community's specific requirements:  high agricultural productivity,
automation of production, automation of the services sector, use of low-grade
ores and production of substitute products, pollution control, and recycling
of scarce materials.  In the present state of technology all these requirements can be met only by a high level of energy consumption.

There are three further reasons for urgently undertaking extensive research and
studies in the energy sectors:

(d)  the fact, that according to past experience, for a new energy source the lead
time elapsing from the design stage through the model stage to large-scale
exploitation is extremely long, approximately 15 years in the most favourable cases or otherwise 25 to 30 years;

(e)  the need to overcome ancillary problems, such as the deterioration of the
countryside and the human environment by the extraction, production, transport, or large-scale conversion of energy;

(f)  the need to find ways of allaying, by demonstration and information, public
concern about the relative dangers, which the layman cannot assess, of the
consequences of an energy shortage, of pollution by hydrocarbons and waste,
and of the risk of accidents in nuclear power stations.

This energy problem continuously emerges as the main physical factor in the minds
of all those who feel committed to the future.  CERD provides good evidence of this
concern.  A specialized working party has been entrusted to Professor Della Porta.
The first chairman of CERD, Professor Casimir, has demonstrated some of the risks

to which a civilization based on excessive consumption would be exposed.  Professor
Meyer, a member of CERD, has been commissioned to carry out a study on a model for
a low-energy-consumption society.  Its results can be compared with those reported
by the US Commission on Energy Conservation:  Figs. 31 and 32 show the results.
They indicate that average per capita consumption and total consumption are increasing at a greater or lesser rate depending on the hypotheses adopted, even in cases
where there is a strong political resolve to make maximum savings in energy consumption.

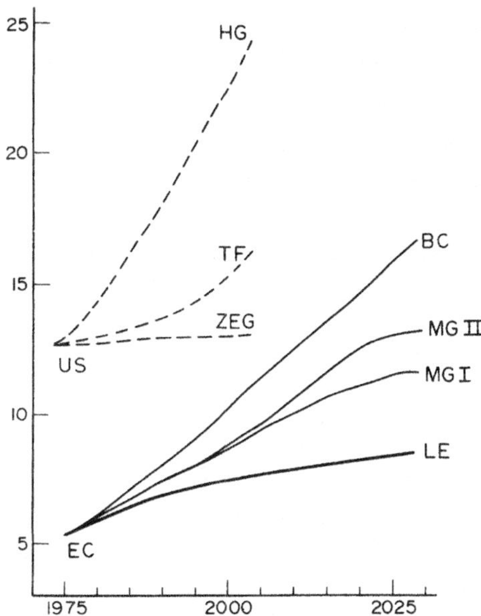


Fig. 31.  Scenarios for the development of the
          per capita energy demand in the EC,
          1973-2025, and the US, 1973-2000. EC:
          base case scenario. MGI, moderate growth
          scenario I. MG II, moderate growth scenario II
          (with stabilizing population). LE, low-energy scenario.
          US: HG, historical growth scenario.  TF, technical
          fix scenario.  ZEG, zero energy growth scenario.

These conclusions show that with the knowledge we have at present and with our
current outlook we are unable to devise a society which would be highly productive
in agriculture, industry, and services and at the same time have a low energy consumption.

In 1971 the average annual per capita energy consumption was 11.5 tce in the United
States and 4.7 tce[1] in Europe.  It has increased continuously since that time,
although there was a substantial slowdown after the 1973 events.  It would be useful to study the model of a society that would preserve the existing standard of
living while reducing average energy consumption in the industrialized countries
to 5 tce per inhabitant per annum.  At the present time, however, we are unable to
imagine practical ways and means of achieving this result.

---

[1] tce = tonnes coal equivalent

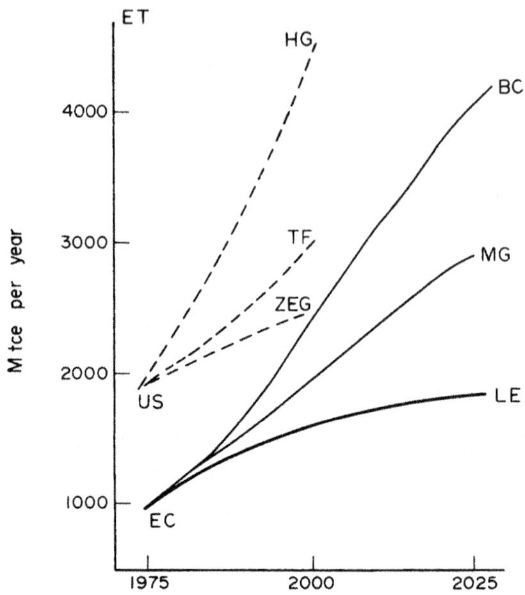



Fig. 32.   Scenarios for the development of total
           energy demand in EC, 1973–2025, and in
           the US, 1973–2000. EC: BC, base case
           scenario. MG, moderate growth scenario.
           LE, low-energy scenario. US: HG, historical
           growth scenario. TF, technical fix scenario.
           ZEG, zero energy growth scenario.

To establish a genuine low-energy-consumption society it would probably be necessary
to go through a stage of adjustment to new consumption models by means of a shift
from quantity to quality, a step which is bound to be progressive and which implies
the existence of a quality supply to replace a quantity supply.  This problem is
tackled elsewhere.

If priority is given to energy production, it is necessary to step up the effort
devoted to all types of new sources, more particularly nuclear energy with special
attention to safety.[1] The greatest danger would lie in postponing full-scale
experiments in the nuclear programmes then finding after several years that economic
activity was slowing down for lack of energy.  This would be followed by rapidly
trying to make up lost ground under the pressure of public opinion by suddenly
speeding up the installation of nuclear power stations without adequate preparation.
If it has an insufficient mastery of the problems connected with the production of
nuclear electricity, Europe might well find itself crippled or completely dependent
on experiments conducted by the United States.  Unlike Europe or Japan, the United
States has no driving need to master nuclear energy as it has a wealth of fossil
fuels, in particular coal and bituminous schists, and is able to rely on political,
and possibly military, power to control oil sources outside its frontiers.  Nevertheless, it is clearly in President Carter's interest to take decisions aimed at
energy conservation and the accelerated development of new sources.

---

[1] Although it appears contrary to common sense, the real safety problems are only
finally solved once the operating stage is reached, as only then is the full extent
of the difficulties revealed.  This does not, of course, mean that all the prepara-
tory precautions should not be studied and, in the case of nuclear energy in particular, prior consideration given to the risk of plutonium proliferation.

## (2)  <u>To increase the productivity of services</u>

The post-industrial civilization is a service civilization.  We have already seen
(pp 59 and 60) that jobs in the service and data-processing sectors will shortly
account for more than half the working population in America.  If the total productivity of the industrialized nations is to be improved, two lines of research must
be carried out simultaneously:

(a)  A study to gain an understanding of the origins of these phenomena and as-
     certain whether it is necessary for so many men to work without ever touching
     objects which will be produced or consumed.  Is this a normal situation in
     a post-industrial civilization or is it an unrestrained eruption of complex-
     ity?

(b)  Regardless of the results of this study, continuation of research and de-
     velopment and the organization of demonstrations on the simplification and
     automation of office working,[1] local and remote data processing, etc.

There are very good prospects of a huge future market for these new techniques
which come within the family of information technologies.  Either Europe will design and produce them and become a major exporter, or it will have to import them
and its balance of payments will suffer greatly, or it will reject this trend for
fear of the social consequences, and overall it will then become less productive
than its main rivals.  How will it be able to live?

This introduction of new techniques to increase productivity in the services sector will have to be accompanied by appropriate studies to ensure that we are not
caught at a loss by the social problems they will cause.  Massive redundancies far
exceeding previous estimates are to be expected in the next twenty years; it is
therefore necessary to devise alternative activities or plan the organization of
a life in which work will not play its usual part.  This brings us back to the
priority that must be given to the question of jobs and unemployment.

## (3)  <u>To increase the competitiveness of agriculture and industry</u>

No attempt will be made to describe all the efforts to be made sector by sector;
we shall merely consider the criteria resulting from the identity of Western
Europe.

It emerges from what has been said above that Community agriculture and industry
should excel in all applications of scientific or technical knowledge where one or
more of the following factors are relevant:

(a)  limited space available; high population density;
(b)  savings in energy and raw materials;
(c)  protection of man and nature against pollution;
(d)  high cost of labour.

A study should be carried out on the application of these criteria, which are more
specific to Europe than to the United States and the Soviet Union; in this respect
Europe has much in common with Japan, normally its main rival.

As an example, the following special fields could be related to <u>criterion (a):</u>

> High-speed cheap transport, both urban and inter-city, over short and
> medium distances.

---

[1]  Burotics: neologism meaning automation of office working.

Intensive crops (agricultural production with a high yield per unit of area
cultivated rather than per person employed.

Criterion (b) calls for:

A high level of development in the recycling industries (water, non-ferrous
metals, organic substances, etc.).

Possible specialization in durable and repairable industrial products.

Heavily insulated heat-treatment units controlled to be operated on
minimum waste cycles:

Production from low-grade ores.

Substitution products.

Criterion (c)  calls for an industry specializing in:

Pollution control (urban, motor vehicle, industrial).

Noise control.

Water regeneration.

Criterion (d)  necessitates considerable effort:

To automate industrial production with the gradual elimination of all human tasks that are arduous, dangerous, or without job satisfaction (automation, robotics, precision engineering, etc.).

To organize work so as to minimize arduous or dirty work (e.g., sorting
and packaging of household waste, non-woven fabrics, etc.).

To establish highly productive industrial units of small dimensions (300-
400 people) where man-machine relations would allow account to be taken
of initiative and interest.

To experiment with methods for the decentralization of power and participation in decision-making.

The launching of special projects of this type calls for a major research and
development effort, a technical standardization policy, and in some cases the adoption of a policy to restrict imports.  For example, it would be unacceptable to ban
a European paper mill or cement works from discharging waste into the atmosphere
or into water and at the same time to expect it to compete with an outside pro-
ducer not subject to the same restrictions.

It is clear that the application of these criteria to research and development
objectives should be preceded by an overall study of the systems analysis type,
leading to proposals for scientific and technical action programmes and the preparation of ancillary measures, in particular as regards standardization.  There would
be nothing more dangerous than to act on emotional grounds without first predicting
the direct and indirect consequences of the changes.

*A Better Life*

For a better life it is first necessary to regain confidence in the future, to re-
vive the hope of improvement, and to learn again that mankind can progress with
all men jointly responsible.  In practical terms, this aspiration to progress is
reflected in development objectives which, as we have seen, can no longer be based
on the growth of material goods for the peoples of the industrialized world.

In our study of the European identity and its interdependence with the identities
of other regions, we believe we have detected three main poles of development:

(a) Development of trade with countries in the process of industrialization.

(b) Improvement of communications between men, dependent on the development of
    the information technologies.

(c) Development of the personality of the individual of small groups by the
    growth of "productive free-time activities".

How can scientific research help to promote these three poles of development?


## (1)  The development of trade with countries in the process of industrialization

Scientific and technical research can be no more than one of the many ancillary
instruments in a major political venture of this kind.  But however modest its
relative position may be, its contribution is far from negligible.

In preparation for decisions, mathematical models could be used to help understand
the problems arising and for co-ordination between the partners.  In the early implementation stages, statistics and model construction would play an important
information role for the partners.  New socio-economic indicators would have to
be devised.

A number of subjects for applied research, on which little is being done today, would
emerge from the development of the tropical countries.  They relate in particular
to agriculture and nutrition, nature conservation, new town planning models, solar
energy, etc.  Other research topics include the transfer of technology, project
studies, the testing of prototypes for small-scale manufacturing units to avoid
spreading the disadvantages of mammoth industry, etc.  New educational and vocational
training methods should be proposed.

Socio-economic research would be necessary to avoid as far as possible falling into
the trap of solutions that would be interpreted as, or felt to be, forms of neo-
colonialism.  Long and difficult studies will be necessary in order to propose systems having free and reciprocal links between the partners, but surely the end is
exciting enough to ensure that the obstacles that appear insurmountable today
will gradually be removed.


## (2)  The development of information technologies

In this study there has been much talk of the productivity of industry and services.  The new information technologies are at the root of these increases in
productivity.  Later in the text there will be frequent reference to "productive
free-time activities".  These can be developed only if a high level of culture
and education can be made universal; this will be possible only by extensive use
of the various new information technologies.  The transfer of know-how to the

developing countries will be largely based on communication techniques.  Information may therefore be likened to the intersection at which all roads meet.

There are three particular areas in which research and development should be carried out.  These are given below in increasing order of need:

  (a)  Activities concerning the support industries proper (telecommunications, electronics, computing, etc.)[1]

  (b)  Activities relating to the applications of the new technologies in various disciplines (medicine, law, administration, linguistics, burotics, robotics, aids for architecture).

  (c)  Activities helping to predict the social and economic consequences of applying these new technologies (effects on life style, distribution of power, personal freedom, civil liberty, working hours, working conditions, education, etc.).

The European Community must take steps to harmonize legislation on this subject (to avoid, for example, the creation of computer havens[2] in the same way as there are tax havens or to prevent excessive disparities in the provisions to protect private life, etc.) and to introduce standardization, especially as regards telecommunications, etc.  It should also ensure that no excessive inequalities arise in the development of the applications of the information technologies, as this might result in the emergence of new underdeveloped areas.

The outlay by the various Member States and the resources available to the Commission of the European Communities to prepare for the future of the information technologies are totally insufficient; all the experts agree on this point.  As regards highly integrated electronic components and microprocessors which it is unanimously agreed will be the nerve cells of tomorrow's society — it is by no means certain that a European company will figure in the five to seven companies in the world in which their production will be concentrated.

In many cases, governments have been deluded by the money they have lavished on manufacturers engaged in an almost hopeless struggle (with trade positions and the state of the art as they are present) against the stranglehold of the most powerful American multinational group.[3] These sacrifices have concealed the inadequacy of the overall R&D effort, especially with regard to small-scale equipment and applications holding out very much greater prospects of success.

In the context of the information technologies and their applications, Europe emerges as a curious mixture of competence, partial success, and underdevelopment.

## (3)  The development of productive free-time activities

It may seem a curious paradox to draw attention to the risk of impoverishment and at the same time to urge a study of free-time activities.  Are we rich enough to plan to reduce working hours?

We have already given a list of free-time activities. (see p. 56).  Each line on this list calls for careful study, if not actual research, to resolve such questions

---

[1] See page 61:  Support industries.

[2] Geographical areas where the excessive concentration of information would be uncontrolled.

[3] IBM — see 1977 OECD study.

as the transmission of knowledge, the necessary cultural level, and educational
facilities.

A number of studies will be required to prepare for the increase in free time and
ways of using it, as it could be very dangerous to declare this as an aim of civilization without a theoretical understanding of the consequences, a forward calculation of their economic implications, and a comparison of experiments carried
out on a small scale. However, as we have already said, prudence counsels daring
in this respect.

*  *  *

This list of priorities does not coincide with the priorities actually included
today in national and Community research programmes.

There is not, however, an excessive gap between them if the armaments development
effort is disregarded. For Europe, it seems right to recommend heading straight
for the goal and acknowledging that the future will depend on economic confrontation and social hope. Consequently the major part of the resources available must
be devoted to these objectives. However, it would probably be too easy, and somewhat demagogic, for scientists to express their views formally on the armaments
policy.

At this point in our thinking, we must ask two further questions:  Is it really
possible to channel the innovation effort and to designate it aims?  If science
and technology cannot be planned, what good is a strategy of priorities?

### CATALYSIS OF RESEARCH AND INNOVATION

At the back of all minds are the failures of, or the enormous cost to be paid for,
some deliberate operations to channel scientific progress and define development
aims.  This experience counsels prudence.  On the other hand, there are enough
spectacular successes, such as the conquest of space, to give us encouragement.
How can we understand this question today?

*Some General Points*

In the past twenty years many competent specialists and research groups have expressed their conclusions on the management of scientific and technical research.
Everything possible seems to have been said.  It is therefore rather discouraging
to have to revert to this question, but for the non-specialist reader who will
have to make up his own mind — either because of the pressure exerted by public
opinion on decisions or because he is committed by a general or specific
responsibility- it appears advisable to outline a few main ideas at this stage.

(1)  An essential part of scientific research meets a deep-felt need of man, a need
     to know and to understand, and it is not necessary, or even desirable, to
     attribute any other purpose to it.  This basic or fundamental research has no
     direct economic justification; it is not motivated by a social, still less
     commercial, objective.  However, it is the vital source of knowledge for ap-
     plied research; it provides a basis for evaluating education; it trains spe-
     cialists who will then be capable of participating in the innovation process.

     As has already been said, this basic research forms the seedbed without which
     intellectual and cultural development could not flourish; it is fertile only
     in an atmosphere of freedom.  The financial resources required are small in

comparison to the other phases of the innovation process.  However impoverished
Europe may be, everything possible should be done to preserve its basic research potential.

(2)  In order to transform basic research results into concepts or products that
     will be useful in satisfying socio-economic requirements, it is necessary to
     create favourable conditions for the innovation process.  This is an extremely
     complex process; in a cycle complicated by continuous feedbacks, it runs
     through the stages of applied and applications research, development, demon-
     stration, and promotion.  It is the process that is normally called R&D but
     we would prefer to call it RDDP, so as to lay more emphasis on the final and
     essential stages of Demonstration and Promotion.

     This innovation process is obviously costly.  It is multidisciplinary and it
     is justified by the socio-economic aspect of the objectives.  The decision to
     undertake an innovation operation must be taken at a high level in a company,
     administration, or state, as the gamble is always considerable and necessitates
     the investment of substantial financial, material, and human resources in re-
     lation to the size of the operative units.

     In the final analysis the factor on which success most depends is the size
     of the potential market for the results and the prospect of disseminating them
     on a wide scale.  This market may be found in the consumer public but it may
     also be artificially created, as in the case of the American and Soviet space
     or armament programmes.  One of the great weaknesses of the nine member coun-
     tries is the fact that the Community markets are not sufficiently receptive to
     innovation.

(3)  Research, whether basic or applied, can be likened to the process of fertili-
     zation.  How much pollen is wasted for every flower fertilized?  How much trial
     and error before a goal is reached?  How many attempts that are bound to be
     fruitless?  However, there are various ways of catalysing reactions to increase
     the prospects of fertilization; we shall come back later to the question of
     the catalysis of communications in scientific and technical effort.  Whatever
     the system, however, the main source of efficiency is, and will remain, the
     intellectual and human quality of the research workers.  Research must be
     carried on by an elite that is highly competent, imaginative, and creative.
     This elitism may be shocking in what purports to be a mass civilization, but
     it is a fact that must be accepted.  The European nations should do all they
     can to encourage their most gifted men to spend a few years of their careers
     in the laboratory.

(4)  The ideas that underlie the proposal of a research programme or that help to
     solve the problems raised are like perishable foodstuffs.  They are useful for
     only a short period of time.  Put into cold storage, they lose their freshness
     or are taken up by rival teams.  Originality will not wait; it calls for rapid
     decision-making processes.  Originality is the reverse of the "scientific
     method", it is not popular with committees and research officials.  And yet it
     is originality that must be preserved.

(5)  The part played by the human element in innovating research, the importance
     of giving researchers some freedom of initiative and allowing them to move
     from one unit or specialized subject to another — all these significant psy-
     chological factors are just beginning to be understood.  The traditional and
     rigid hierarchical structures of European research must be replaced by par-
     ticipative methods.  The tendency to immure research workers on our old con-
     tinent in protective and immovable structures must be counteracted by the mo-
     bility of young scientists.  This problem is far from being solved.

*The Innovating Effort can be Guided by Catalysis*

From all that has been said above it is clear that there can be no authoritarian
planning of research objectives and that it is necessary to allow a degree of freedom and spontaneity.  That is why moderate methods of intervention, described here
as research catalysis, have been devised and gradually tried out.

What is research catalysis and by what means can catalytic action be applied and
controlled?

(1)  The first form is the establishment of communications.  Since the beginning
     of history, scientists have been great travellers; they have a constant need
     to assess their results and the lines of their research for themselves by
     comparing them with those of other experts working at a distance from them
     or in a different environment.  Reciprocal visits, symposia, congresses, sem-
     inars, publications, and documentation centres are the traditional instruments
     for communication.  This self-assessment also leads to self-co-ordination.
     The scientific world has no frontiers; as soon as scientists or technologists
     experience a mutual esteem, a desire to work together is born - except in the
     reverse case where emulation becomes aggressive rivalry, but this is equally
     effective in providing strong motivation.

     Within a single discipline, experts do not generally need outside agents to
     organize meetings.  They know whom they have to see, when, where, and on what
     subjects.  But often in Europe there is a shortage of funds for travel, enter-
     tainment, or publication, and in any case the slender resources tend to be
     used for visits to American rather than to European colleagues.  An organiza-
     tion which could obtain some extra money to subsidize this travel, to award
     grants for stays abroad, or to organize meetings would have an appreciable
     catalytic influence.

     This influence would become catalytic guidance if the meetings subsidized were
     of a different type from those that would otherwise be organized spontaneously.
     For example, meetings between European experts from different countries rather
     than bilateral meetings with the United States; or the organization of a
     multidisciplinary seminar bringing together participants who would normally
     be unknown to each other to discuss a topical subject; for example, scientists
     from basic and applied research, manufacturers, and potential consumers.  Those
     who are accustomed to managing catalytic systems of this kind have learnt from
     experience that the contacts made between participants are not broken off when
     the meeting ends; they continue and lead spontaneously to bilateral or multi-
     lateral co-operation between partners who would never have known each other if
     the original event had not taken place.

(2)  A more elaborate form of catalytic action specific to applied research and
     RDDP consists of carrying out studies on paper, possibly with the assistance
     of mathematical simulation, so as to predict the relevance of certain objec-
     tives, calculate their direct and indirect effects, and describe the resources
     and structures that would have to be provided to attain them.

     These studies involve talks between possible partners coming from different
     specialized fields or having different interests.  They attract interest and
     pave the way for the consensus that will be necessary for the establishment
     of task forces[1] and "laboratories without walls" that are so important for

---

[1] Task forces:  these are, so to speak, research "commando units" bringing
together carefully selected specialists, the team being dissolved once the aim
is achieved.

the cross-fertilization of multidisciplinary research.

It must be realized, however, that the studies are valuable only if they prepare the way for decisions on action.  If they come to nothing because funds
are not available, or because decision-making processes are too long or so
complex that a negative response is all too likely, these studies generate a
feeling of futility and frustration.  It must be admitted here that the
Commission of the European Communities, in most of its Directorates-General,
provides an example of intense activity in the way of paperwork issuing from
numerous committees and working parties.  Although thy reach excellent conclusions, these are never followed up or are followed up too late, so that
the results are out-of-date.  This might be called anti-catalysis.

Although in some sectors, such as common financial or industrial policy,
Europe is so divided that it is impossible to remedy this situation, it is
unthinkable to allow scientific research to become bogged down in the same
way.  As will be seen later, arguments such as "a fair return" or "supra-
nationality" should not seriously be allowed to prevent a European executive
body for science and technology being given the power to make financial and
other commitments, within the limits of a specific budget, when initial
studies show that certain useful objectives can be successfully attained.

(3)   A further step is taken in catalytic intervention when a body has funds to
      finance outside research.  Generally speaking, a very little money is suf-
      ficient to influence research orientations; to what is this phenomenon due?

      In the case of research centres belonging to private companies or non-profit-
      making organizations, there is generally a shortage of funds for the opera-
      tion of the laboratories; a customer placing research contracts is always
      welcome.  If the customer requires an internal effort equal to his financial
      participation, and if the subject concerned is not outside the field in which
      the company specializes, the contract is readily concluded.

      A research centre belonging to a public body is always slightly or even ex-
      tremely short of money to purchase special equipment, for missions, for pub-
      lication or documentation.  Consequently the laboratories are ready to accept
      outside resources in the form of contracts and they acknowledge that in such
      cases they have to follow the wishes of the person putting up the money.

      In this form of catalysis by research contracts it is possible to organize
      co-operation in the study of a single subject between several laboratories in
      different countries, specializing in different subjects and having different
      natures (public, semi-public, non-profit-making associations, research centres
      in private industry).  This provides an excellent method of disseminating
      information, initiating co-operation, and promoting redeployment on a Com-
      munity scale.

      These contracts often have one shortcoming however:  they are subject only to
      what might be called a judgement of peers based on the publications put out
      after the research by the team working on the contract.  Consequently the per-
      son putting up the funds often finds it difficult to assess the results, as
      all he obtains for his money is paperwork or experimental results, and he is
      not easily able to promote exploitation by the economic and social sector.

(4)   There is one method of intervention that is a more powerful spur and is
      judged on an all or nothing basis.  A typical example of such programmes is the
      Apollo operation.  The aim is to carry out, within a specific period and bud-
      get ceiling, an ambitious objective making wide use of new techniques and

expected to yield extensive spin-off in many allied fields.

The armaments and space programmes have proved to the United States the validity of its role as a driving force in applied research.  The U.S. examples cannot be carried over into Europe but they can serve as the pattern for a policy of catalytic activities organized around medium-sized pilot projects whose success or failure would be simple to check in the course of the project.

### Contracts for the Exploitation of Results

When a European laboratory obtains a result that would appear to be of practical use in the economic sector, industry is often reluctant to finance its development. Company managements often lack the technical background to grasp the practical advantages to be derived from the results or to assess the prospects of success and calculate their risks, or they may quite simply have no confidence in the research workers with whom they are dealing.  Sometimes the invention is taken up by the more receptive and aggressive U.S. market and European manufacturers then purchase a licence to work the patent and the necessary know-how without even realizing that the invention was conceived in laboratories close to their factories.

That is why it is often useful for an intermediate body that has followed the research work to propose to a manufacturer that it will finance part of the development effort.  Once the first link-up is made, collaboration will follow smoothly and automatically.  Naturally, the earlier co-operation has been organized between the laboratories and the "development department" or "test site", the more smoothly things will run; as far as possible contact should be made as soon as there is a prospect of obtaining results that have some practical utility.

This type of intervention should be distinguished from that known as development aid.  This consists of support from an outside body for a project study or demonstration project that a manufacturer wishes to carry out to launch a new production method or a new product.  Development aid may be regarded as coming under the heading of industrial policy.

Exploitation aid is something that extends the research project to make it of practical use:  it comes under "scientific and technical policy".

### Means for a Policy

The above considerations on the characteristics of scientific and technical research show that it is possible to influence research objectives and define ways and means of implementing a policy.  This will be effective only if the following conditions are fulfilled:

(a)   there is in existence strong, free basic research not excessively worried about its future;[1]

(b)   there are extensive facilities for innovating research.  These facilities must be such that they establish a continuous flow of exchanges, with numerous feedbacks, between applied research, applications research, development demonstration, and promotion;

---

[1] This raises the whole question of the career of specialized research scientists working full time (not examined here but a particular problem in Western Europe).

(c)   there is a judicious amount of interpenetration between basic research and
      innovating research;

(d)   there are bodies acting as catalysts to improve efficiency and provide guidance.

The complexity and subtlety now become clear.  High-quality basic research can
be obtained only from a "contract of confidence", as the indispensable freedom
makes it impossible to intervene directly in the choice of subjects.  The initiative of the commercial world is also the expression of a freedom:  right to originality, right to take a chance and accept a risk, right to intervene outside
the area of the planned objectives.  The pursuing of objectives of general interest,
on the other hand, reflects a deliberate policy of planning scientific and technical
effort.

This counterbalance of planning and freedom must be respected.  To ignore it would
be to lose the benefit of originality which is essentially unpredictable in nature
and in the order in which it enters the sequence of events.  This brings us back
to one of the main ideas expressed earlier.  Research is a process of fertilization,
its prospects of flourishing must be left to chance; chance is of a very special
nature here since it is backed up by the inventive minds of research scientists.

This subtle complexity brings us to the question of the ownership of the resources
connected with the planning objectives.  Research resources consist above all of
teams of research scientists and technicians working together in laboratories or
scattered throughout administrations and industry.  These research scientists
have a specialized subject to which they are strongly attached, as they have put
enormous effort into mastering it, obtaining contacts, and making themselves
known.  These research scientists are also strongly attached to their discipline
as it is not difficult for them to let themselves speculate and to see the justification for their effort in the successful results that they expect to attain.
These feelings of attachment to a profession, praiseworthy in themselves in many
respects, have the effect of crystallizing the future on the pattern of the present.  There is little, and in some cases no, mobility of objectives, still less
mobility by changing specialization.  Where these resources belong to a body that
is powerful or not very mobile — such as a major industrial group, a government
sector, an international organization — these stability factors are all the stronger
and there is less possibility of moving towards new objectives or of encouraging
originality to flourish.  These comments account for the occasional difficulties
experienced by industrial groups with their central laboratories and for the fact
that, in research done in state laboratories, at least 95% of the budget appropriations go to the continuation of existing work, leaving a maximum of 5% — plus
growth appropriations — to respond to changes required by the economic and social
environment.  This is one of the important paradoxes of research that must be borne
in mind:  the possession of expensive facilities does not give freedom but tends
to confine research to its previous lines.  Thus not only is research working for
the long term but it is marked by considerable time constants.

The following advice can be derived from this analysis:

(a)   The freedom so greatly needed by research must be guaranteed by ensuring a
      large number of financing sources.  Unless a laboratory is closely integrated
      in an industry, it is better for it to be funded jointly by its owner and
      by outside financing agencies of a national and international nature.

(b)   Research budgets must be multi-annual;  they must predict lines of development
      sufficiently far in advance to allow researchers to prepare for them.  They
      must be protected against economic and cyclical fluctuations.

(c) When innovation objectives are defined in relation to the public interest
    they must as far as possible be entrusted to contractors independent of the
    persons carrying out the research, which implies that the contractors are not
    themselves the owners of the research facilities.  They will merely be respon-
    sible — a considerable task — for finding the appropriate skills when the time
    comes, for putting them to work, and for co-ordinating them.

We shall come back to this advice when we study ways of implementing a policy at
Community level.  The reader will already have noted one of the main conclusions:
it is necessary to promote flexible, active, and responsible catalysis that is
only loosely interconnected with national policies.


## NATIONAL STRATEGIES AND COMMUNITY STRATEGY

All that has just been said about the necessary freedom of scientific research
and innovation argues in favour of a wide variety of financing sources, without
very close co-ordinating links between them, and of a wide variety of participants.

All too often large research concentrations are inefficient, difficult to control,
overburdened by costs, and do not encourage originality.  If there is one field
in which power should be distributed and generously shared out, it is scientific
research.

Hence there is no occasion to regret too keenly that national and Community policies
within the Europe of the Nine are developing on somewhat divergent lines.  If for
a special reason such as a national bent or confidence in a particular specialization, a Community country wishes to make a major effort in a sector regarded as
non-priority in the light of Community criteria, by what right should it be prevented?

This remark does not by any means imply that there should not be a Community policy;
on the contrary, the efforts towards it should be greatly stepped up as the current
level is totally inadequate, but the objectives must be specific to the interests
of the Community.  Going beyond national policies, the diversity of which must be
maintained as a very precious asset, this policy will be an expression of Community
solidarity; it would embody in its strategy the will to construct a Community.

Nor does this comment imply that it is not desirable to organize synergy between
European researchers wherever they may be working.  On the contrary, it is essential
to stimulate co-ordination by research teams themselves, to disseminate knowledge
and to avoid duplication.  However, in developing synergic effects, reliance must
not be placed on co-ordinating national programmes by tackling the problem from
the top; it must be attacked from the base by catalysing communications between
researchers and carrying out joint projects.

In addition, it is necessary to catalyse the transformation of industrial and
economic circles to make them receptive to innovation.  It must never be forgotten
that in the United States the rich harvest from innovation is due much more to the
fertility of the environment than to the relevance of scientific policies.

Needless to say, there will be many instances where the scale of the facilities
required for research, and even more for development, will be beyond the capacity —
in terms of economics or manpower — of any one Member State.  There should then be
no hesitation in setting up a structure adequate for the scale of operations required (e.g. in space technology, nuclear fusion, high-energy physics, etc.).  It
would seem that, in the past, although these situations have been well enough understood, there has been too little resolution in the decision-making process and

not enough delegation of authority in the execution.

*The Background to the Community Research Policy*

It is interesting to compare these general ideas with the facts observed when the past history of Community R&D is examined.

Over the years, the Community scientific research policy has been built up layer by layer. The ECSC (European Coal and Steel Community) Treaty (1952) and Euratom or EAEC (European Atomic Energy Community Treaty (1958) made specific provision for common research and development facilities. It would appear that the guiding idea at that time was the concentration of facilities in order to profit more easily from the scale effect resulting from the size of the Community.

The Rome Treaty which established the European Economic Community, has no specific provisions except in the case of agricultural research (Article 41 of the EEC Treaty). In the mid-1960s the Euratom aims proved unattainable against the political background at the time. Coal and steel research are aided by financing sources provided by the ECSC; agricultural research is a special item and there are no action programmes for the other topics. The most that can be said is that Research Europe is not yet under way.

In 1967 a PREST (Scientific and Technical Research Policy) report proposed measures for the formulation of a Community scientific and technical policy. It would appear that at that time the predominant idea was to co-ordinate national policies.

The desire for action was re-stimulated by the Paris Summit meeting in 1972. The Heads of State and Government came out in favour of the development of a common policy in the field of science and technology which would require "the co-ordination of national policies" and "joint implementation of projects of interest to the Community".

This decision was put to practical effect by a number of decisions taken by the Council of Ministers in January 1974, which laid the first foundations for the definition and implementation of a common policy in the field of science and technology.

In the years that followed, several steps were taken:

(a)   the setting up of CREST (Scientific and Technical Research Committee), consisting of senior national officials and chaired by the Director General for Research, Science and Education;[1]

(b)   the setting up of new ACPMs (Advisory Committees on Programme Management);

(c)   the setting up of an advisory committee consisting of twenty-one independent individuals; this was CERD (European Committee for Research and Development);

(d)   the establishment of systematic links with the European Science Foundation[2] founded in 1974;

---

[1]   Directorate-General XII of the Commission of the European Communities

[2]   The ESF consists of forty scientific organizations in sixteen European countries, including the nine Community countries.

(e)  the presentation of an initial communication of R&D aims and priorities to
     the Council of the Community (November 1975);

(f)  an organizational restructuring and the establishment of a new programme for
     the Community's Joint Research Centre (JRC), which brings together, under a
     single authority, the laboratories established under Euratom (Ispra, Geel,
     Karlsruhe, Petten);

(g)  the development of research on nuclear fusion and radiation protection and at
     the same time the preparation of the project for an experimental fusion reac-
     tor: the Joint European Torus (JET);

(h)  the forwarding to the Council by the Commission, in June 1977, of the guide-
     lines for common policy in the field of science and technology (1977–80).
     These guidelines form the framework within which Community projects will be
     defined.  They lay down priority sectors and selection criteria used to de-
     cide on Community projects conducted in the form of either "direct action"
     in Community centres, "indirect action" by outside contract holders, or
     "concerted action" projects in which the Community merely ensures co-ordination
     and disseminates information without providing any funds for the laboratories
     (Table 11)

Table 11.  Community R&D programmes

Funds allocated up to the end of April 1978 (million u.a./EUA)

| Sectors | Joint Research Centre (dir.action) | Contract Research (indir. action) | Co-ordination (concerted action) | Period |
|---|---|---|---|---|
| **1. Energy** | | | | |
| Nuclear fission | 150.70 | 62.66 | None | 77–80 (with some residue from 75–76) |
| Nuclear fusion | 12.75 | 124.00 | None | 76–80 |
| JET | – | 147.68 | None | 77–83 |
| New forms of energy | 41.29 | 47.62 | None | mid-75 to mid-79 and 77–80 |
| Energy-saving | – | 11.38 | None | mid-75 to mid-79 |
| **2. Industrial policy** | – | 4.59 | None | 1976–81 |
| **3. Environment** | 29.95 | 16.00 | 0.14 | 1977–80 (with residue 1976) |
| **4. Raw materials** | 7.49 | 23.90 | None | 1977–80 |
| **5. Agricultural policy** | – | 10.30 | 0.25 | 1975–80 |
| **6. Medical and social** | – | – | 1.29 | 1978–81 |
| **7. Long-term R&D policy** | token entry | token entry | token entry | 1977–80 |
| **8. Miscellaneous** (training, measure-ments, etc..,) | 154.96 | – | None | 1977–80 |

Table 11 is a synopsis of an internal document of the Commission of the European Communities entitled "Community R&D programmes — situation at the end of April 1978".

Since the outcome is still unknown, these figures do not include the proposals which will be presented to the decision-making bodies of the Community at the end of 1978 (reactor safety, decommissioning of reactors, data-processing, FAST programme, etc.).

It should be remembered that the research managed by the Community is distinct from that carried out within the ECSC framework.  Moreover, most of the Directorates-General have study appropriations to which no reference has been made here.

The extent of the resources may be assessed by reference to 1976.  Budget appropriations for Community research amounted to 162.8 million units of account or $180.7 million.  This corresponds approximately to 1.2% of the total publicly funded research and development effort in the nine Community countries, or 0.6% of the total funds devoted to R&D in the Community, allowing for expenditure by private undertakings, which was of the same order as that of public bodies.  Of these funds, approximately 100 million u.a. are taken up by the priority programmes and by the Joint Research Centre.  In the latter the mobility required to adapt to new aims is hampered by the obstacles normally found in large structures, although it has been agreed that this will be given special attention in the drafting of each four-year plan.  Consequently, 80 million u.a. remains for synergic projects, approximately 0.2% of the total European expenditure on research and innovation. This is barely a homeopathic dose, let alone a therapeutic one.

It cannot be denied, however, that encouraging progress has been made even if the scale of the resources is still extraordinarily modest.  The decision-making circuits are in keeping with the Community pattern, being pointlessly cumbersome, and consequently the methods used are still excessively technocratic.[1]


### Desirable Innovations

If Europe's progress towards a new development model is to be promoted gradually and by means of mild therapy, all that has been said above indicates that a number of changes must be made.

Amongst the negative changes, it would be necessary to give up the idea of any organic concentration of research facilities within centres belonging to the Community.  The exception, such as the JRC, is excepted only to confirm the rule. Illusory ideas on the co-ordination of national policies would also have to go. These policies are either virtually non-existent or linked to dominant industrial systems, or governed by national objectives unlikely to be influenced by outside forces.

The positive changes would be based on the general idea that science and innovation are steps towards action in which the protagonists are research scientists and engineers, in contrast to administrative set-ups.  The method proposed for Community intervention would essentially be catalytic in action.  The main types of catalytic action examined in an earlier chapter are summarized in Table 12.

---

[1] This comment does not reflect on the value of the remarkably able, tenacious and open-minded senior Community officials; it is aimed directly at the shackles to restrain their freedom of action imposed by a system of controls and committees.

Table 12. Types of catalytic action on research and innovation

1.  Meetings, symposia, seminars, publications, extension work

2.  Studies, models, futurology

3.  Research contracts

4.  Pilot projects

5.  Exploitation contracts

To conduct a policy of catalytic intervention, encouragement and incentive, it would be necessary to have:  a suitable structure and facilities; a set of reference criteria to show how to select specific fields for action; and a policy for selecting participants.

*Structure and Facilities*

The launching of a programme of catalytic projects, supervision of performance and correction of false moves or errors by the partners resembles in all respects the work of a prime contractor.  It will therefore be necessary to have an <u>executive body</u> with the following characteristics:

(a)  capable of taking rapid decisions (research ideas are perishable foodstuffs that must be consumed fresh);

(b)  responsible to the relevant political authority in advance as regards the content of the programmes and after the event as regards their execution;

(c)  capable of encouraging originality by taking the advice of independent experts;

(d)  having numerous degrees of freedom, in particular with regard to national policies, so as to adhere to the rule of a wide variety of financing sources;

(e)  having a substantial multi-annual budget enabling it to go well beyond the homeopathic doses administered up to now.

This is not the place to define this executive structure.  Once its principle is decided, a serious study would have to be carried out so as to set it up without unduly burdening the Community with new bureaucratic procedures but rather trying to alleviate it and if possible to integrate the Joint Research Centre in a vast system of extra-mural exchanges.  The level of the budgetary appropriations would be a political matter of the first importance.  Now that the way is being prepared for elections to the European Parliament by direct universal suffrage, and in view of the tendencies expressed by public opinion, it would be desirable to devote substantial funds to European science and technology.  The need for a decision to try to catch up on the efforts of our main commercial rivals outside the Community would indicate that these additional appropriations should be granted without a proportionate reduction in research and innovation budgets at national level, which are already inadequate.

*Characteristic Criteria of a Community Policy*

The general idea underlying the new proposals for a Community science and technology

policy is the recognition that for European R&D there are certain factors specific
to the existence of the Community.  These specific factors cannot be taken into
account at national levels, as in the national context their expression would not
be encouraged and they would not be given the priority they merit.

At the Milan symposium (May 1976) certain "Community criteria" were established.
After making the alterations that we deemed necessary, we can describe these
factors specific to the Community under four categories of ideas: dimension,
homogeneity, political consequence, communication.

There is no need to go into the dimension factor at length.  It means that projects
requiring facilities for execution beyond the scale of national facilities should
be transferred to Community level.  This  applies to certain aspects of energy,
space, information technology, the exploitation of marine resources, etc.  It does
not in any way mean that states can no longer continue to work in these fields at
national level.

The homogeneity factor is a fundamental element from the political viewpoint.  If
we wish to develop towards a genuine community, it is unacceptable that the concentration of scientific and technical assets to the benefit of a limited number
of countries or regions should lead to excessive disparities in wealth, working
conditions, industrial competiveness, or welfare.  This applies, therefore, to
everything connected with health, life at work, spare-time activities, environmental
protection, the automation of production and services.

Political consequence may justify special projects in areas where Europe must
present a single image in order to carry weight with her international partners
in negotiations where the technical factor is important.  The standardization of
computer equipment and networks is a good example.  Some aspects of research for
the benefit of the developing countries could gain from being presented under the
Community label.  International law on the exploitation of the oceans is another
example of a political-technical problem.

Communications between scientists can lead to useful savings by stimulating emulation between research workers in different countries, helping to reduce duplication,
speeding up the dissemination of knowledge and innovation, and catalysing self-co-
ordination amongst research workers.[1]


### Partners

A policy of catalytic action is valid only if it applies to numerous free, responsible, mobile, imaginative, and daring partners.  The futility of the environment
in respect of innovation has seriously deteriorated in Europe.  Governments are
sometimes embarrassed by the huge structures they have themselves set up; these
structures have to some extent crystallized as a result of the proliferation of
certain considerations of corporate interests expressed by research scientists
without taking sufficient account of the general interest, and in particular the
needs of the economic sector.

All too often, inertia prevails over mobility, specialized work prevails over
multidisciplinary research, a research scientist's career is based on the intrinsic

---

[1] Care must be taken not to use the wrong methods; these objectives of communication and self-co-ordination will not be attained by comparing programmes, i.e. not
by bureaucracy but by action, by associating various teams of research workers in
a joint study, and carrying out a joint project (partly financed by the Community).

merit of what he publishes rather than on a concern to find practical applications.
The "interstitial" life, that of the non-profit-making societies, foundations, and
undertakings set up by researchers, is not well developed and is tending to become
more and more stifled.

The situation is the reverse of that in the United States, and we must realize this.
Whereas in the United States there are numerous relay points between research and
its practical utilization (non-profit-making foundations, institutes specializing
in certain functions of the transfer of knowledge, independent universities,
private research companies, design companies, industrial research centres, small
and large bodies subsidized by the Federal Government, the States, the large
towns, Federation of Scientific Associations, technical interest groups), Europe
is marked by the rigidity and monotony of its structures.  There is an arterial
research network but almost no capillary network.  And yet initiative, originality,
the beginnings of innovation come from the most part from cells supplied by
capillary action.

If national governments are enmeshed in the complicated structures they have themselves set up, the Community has the huge advantage of being unencumbered by any
heritage of this kind.  Consequently it can by catalytic action help to promote
the creation of this interstitial life, these bodies capable of forming bridges
between basic research and industry.  In this way it would contribute towards the
development of the fertility that is to a great extent lacking.


*Fields Suitable for Intervention*

Apart from the judicious choice of partners for catalytic activities, the main
aim should be to eliminate the vulnerable lines of the European research and
innovation system.  The sensitive areas are situated more particularly at the following meeting points considered individually or in their relations to each other:

(a)   the interface between basic research and applied research;

(b)   the interface between applied research and industrial development;

(c)   the interface between social sciences and exact sciences;

(d)   the interface between human requirements and scientific and technical pro-
      grammes;

(e)   and, of course, the exchange of knowledge and exploitation of results within
      Europe.

Naturally, depending on the case in point, the sensitivity of these vulnerable
areas varies and the examples given above are not restrictive.

*       *       *

Thus a Community strategy and procedures distinct from national strategies and
procedures emerges more and more clearly.  This Community policy is not a supplementary or complementary accessory, which would be of minor importance in
comparison to that of the national states, but is one of the most important expressions of the will of the states to benefit from their solidarity as members
of a Community.  While preserving their own identities and differences, the Member
States could thus overcome the major obstacles which, if they acted in isolation,
would hamper their harmonious technological development and as a result their economic and social development.

# CONCLUSIONS TO PART II

In view of the state of the world and the rivalries being acted out in it, and in view of the potential for progress generated by recent scientific discoveries, there is every sign that the phenomenon of technological evolution has not dried up. On the contrary, its pace will increase in the next thirty years provided it manages to solve the financial problems. The radical changes that have occurred in the space of a hundred years in the way of electrification, public health, agricultural and industrial productivity, road transport, and urbanization are probably less important than those foreshadowed with the economic and social conquest of new territories such as nuclear energy, space, the ocean, bio-techniques, information technologies, and new instruments to improve productivity in industry (robotics) and the services sector (burotics).

To survive, Europe must improve its innovation effort in both quantitative and qualitative terms. From the former angle, if we wish to raise ourselves to the American or Japanese levels, improvements of 50 to 100% must be made. From the qualitative aspect it is necessary both to increase the fertility of the research environment and to supplement the Member States' efforts by Community action, the only way of overcoming a number of vast and heterogeneous obstacles. The theory upheld here is that competition must also be shifted to the area of man's social needs and desires so as to get away from the prestige and power aims where Europe is handicapped and the post-industrial society is taking the wrong turning.

The Community scientific and technical policy should have an important part to play in this effort towards consolidation and redeployment. It must be separate from national policies. It would be specific to European solidarity if it adopted as its priorities the criteria of size, homogeneity, political weight, and communication. In the proposed project, the financial appropriations would be greatly increased but would not be used to develop the bodies under the direct control of the Commission of the European Communities. The co-ordination of national policies would have only limited ambitions, so as to allow Member States to give rein to their diversity and leave them great freedom in the choice of national priorities. On the other hand, a responsible executive body would be set up to organize energetically, by means of flexible and rapid decision making, the implementation of a policy to catalyse scientific research and innovation. The main aim would be to increase the synergism between European research scientists and engineers. This policy of catalytic action would aim at promoting the creation of an interstitial life encouraging innovation, in particular by swelling the funds devoted to the sensitive areas forming the interfaces between basic research, applied research, and industrial development, and by devoting particular attention to the meeting point between the exact and human sciences.  This would be done by organizing meetings between European research scientists, setting them joint objectives under contracts for research or pilot projects carried out by multidisciplinary and multinational teams.  The "laboratories without walls" and "temporary contract supervision" would enable efficiency and mobility to be combined.

To arouse controversy and fire imaginations, a list of priority subjects is proposed that takes account both of certain ascending trends particularly evident in contemporary science and of the objectives of survival and a better life.  This is mild therapy, but it would be wrong to believe that its effects will be negligible, as can be shown by two examples concerning public research systems and industry.

The machinery of research carried out in universities and public bodies has become excessively cumbersome over the past fifty years.  This situation is reminiscent of the excessive training required for Olympic sports at the level of international competition: instead of games played by gifted amateurs, these physical sports have become techniques compelling athletes to accept a way of life and medical care quite out of the ordinary.  The "brain sports", as one might in the past have called the intellectual activities of research workers, a small and elite group, have become forces recruited to fight international competition.

Forced to find an effective reply to the competitive pressure exerted by other countries, Europeans appear to have lost the feeling for the fair balance between the degrees of freedom that have to be granted to these different stages in the acquisition and practical utilization of scientific and technical knowledge.  Innovation is a long climb from basic research to applied research and applications research, then on to development and demonstration.  The higher one moves up the scale towards practical exploitation, the greater must be the pressure for a practical and immediate result, the greater and more precise the facilities, the more administration and management prevail over inspiration, initiative, and imagination. The closer one is to basic research, the more necessary it is to leave some scope for chance, the unexpected, the originality of the researcher's mind, and the more the constraints on freedom must be reduced.

In Europe, all too often, management methods that are of use only in the final stages of innovation have been transferred to the fields of basic and applied research.  As a result, much of the effort to acquire knowledge and much inventive work is today over-administered, over-co-ordinated, and over-controlled.

The university has been the chief victim in this gradual transformation of its vocation and its administration.  It is expected simultaneously to change over from elitist education to mass education, to adapt to the explosion in scientific knowledge, to take on basic research and to establish efficient links with industry for applied research — too much to hope for in one go.  It is necessary to rethink the whole apparatus, probably to set up intermediate bodies having different statutes, to solve the problems involved in the links between research and applications.

This is being done in most of the European countries, but the process encounters numerous psychological inertias.  Those enjoying advantages already acquired are reluctant to relinquish them without an assurance that they will be participating in real progress.  Progress is blocked when an attempt is made to tackle this problem at national level.  Probably a change to a Community basis would, by allowing dissimilar experiments to be compared, reopen the dialogue in a constructive fashion.  This would be one way in which Europeans could take advantage of their diversity in order to help each other to make progress.

Turning to industry, we can imagine for a moment what European company structures would be like if manufacturers, forced to seek alliances in order to attain the necessary scale of operation, had been able to learn by experience to know their European colleagues through research contracts jointly concluded at the initiative of the Community. How many of the industrial mergers that were concluded at national level, often creating national monopolies not really in the general interest, would then have taken place between European manufacturers, thus establishing the foundations for future multinational groups capable of counterbalancing the non-European groups?

* * *

It is clear that the ideas expressed in Part II of this work are far from guaranteeing a solution. They should be taken as rough proposals needed to stimulate constructive discussion. They do not claim to be exhaustive, and many fundamental issues have not been covered. The work may therefore be considered as the beginning of a study rather than a conclusion.

Some members of CERD, influenced in particular by their industrial experience, have expressed the opinion that the solutions proposed in this Part II would not be sufficient to give Europe the innovating capacity it needs. They would like to see four complementary measures added:

(1)   Promotion of the mobility of personnel between research centres everywhere and the economic sector. This mobility should reflect the scientists' awareness of their role in the construction of society.

(2)   The establishment of funds to finance innovation so as to make up for the unwillingness shown by European commercial banks to invest in small — and medium-sized ventures with a high technical risk. This could be the aim of a European Innovation Bank.

(3)   Encouragement for the formation or development of multinational European-based undertakings having as their object investment in North America, the Far East, and the countries in the process of industrialization so as to counterbalance the influence of the major American and Japanese companies.

(4)   The adoption of a standardization policy taking account of the factors specific to Europe, in particular protection of the environment and possibly certain social factors.

This study of these questions has not been pursued, as the subject tends to fall under industrial policy rather than scientific and technical policy.

# TOWARDS THE SECOND RENAISSANCE?

How strange is Europe's situation today!

Europe invented the modern form of scientific and technical research and in the recent past still regarded it as the source of progress. Today, it is allowing its most direct rivals to take the offensive with more powerful and better-employed resources.

Europe invented the university but no longer knows how to make the best use of it. It overburdens it instead of leaving applied research, for economic and social purposes, to others, and yet it cannot manage to break the isolation of the academic research staff who educate its youth.

Europe invented industry which, replacing craft, established the domination of the economy of scale. It still has one of the most concentrated and best-educated bodies of consumers but it lacks the cohesion to provide a unified market for the sensitive products of innovation and the advanced technologies that are paving the way for tomorrow's world.

Europe was the cradle of the artificial world around us which moulds the sensibilities of our children, remote from contact with nature: industrial noise, chemical odours, landscapes of skyscrapers and factories, contact with artificial fabrics and plastics, synthetic food and drink, preventive medical care, mechanical transport, continuous availability of audiovisual entertainment. But Europe is reluctant to step up its efforts on these lines; it is lagging behind in the progress accompanying the advent of the machine kingdom. As a result, the position of superiority that Europe has won for itself in the international division of labour are now being challenged one after the other.

And yet never has the need been expressed in such inarguable terms. Dependent on imports for its survival, Europe can pay for its purchases only by selling manufactured products or capital goods. Intrinsically poor except in manpower, it must now negotiate the raw material and energy resources that it had for centuries been accustomed to control politically in more than half the developed world. Poorly equipped to counter the diffuse violence of the poor peoples, Europe is also on the verge of being confronted with the most serious population imbalance that the history of the world has ever known.

Is Europe on the way to impoverishment?

* * *

The feeling of danger is, however, poorly understood because many factors tend to reassure us for the immediate future:

(1)  Economic patterns are not broken as easily as political treaties.  Despite de-colonization, therefore, Western Europe remains one of the major trading and financial centres of the world; its political stability encourages the invest-ment of capital; some world commodity markets are still centred there.

(2)  If world economic and political balances are examined, it is clear that West-ern Europe's role is still a major one; its eclipse would have unforeseeable consequences.  Neither the Soviet Union nor, above all, the United States show any intention of destroying that balance; on the contrary, they appear anxious to consolidate it.  This attitude is in keeping with history; the hegemonies have always tried to preserve the balance of forces on which they were estab-lished.  All the European options are implicitly based on the postulate of the American guarantee not only in defence but also in economic matters.

(3)  Europe's force today is based partly on the weakness of others:  the low com-petitive capacity of the manufacturing and capital goods industries in the planned-economy countries; the long way to go for the countries in the process of industrialization,  however rich they may be in primary resources before they gain any real power in international trade in the products in which the Europeans excel.  Even Japan is itself threatened by the excesses stemming from its success:  saturation of building land; pollution; extreme dependence on commodity imports; risk of a revival of protectionism by its customers.

(4)  Europe has built-in assets resulting from its temperate climate, its domesti-cated nature, the strength of its investments, and still more from the quality of its peoples accustomed to industry and international trade and the wealth of its cultural background.  It is also an ideal partner for countries eager to continue their development or to work towards their economic take-off because it now combines competence with an absence of imperialist ambition.

This combination of favourable factors guarantees Europe's survival, and perhaps even relatively comfortable existence, for a period that is difficult to predict but is probably no more than a few decades.  These balances are precarious; some are no more than the relics of a past that will be eclipsed, others depend on external forces.  What would happen if American public opinion grew tired of bear-ing the weight of the world or if North American technological superiority were no longer able to contain the consequences of the population growth in the south?  Can we take the risk of entrusting the fate of a continent to outsiders?

But let us keep calm there is still some time in which to act.

The gravity of the European situation is reflected in the abnormally deep-rooted conflict between short-term and long-term interests.  If the Europeans make no sacrifice for the future, things will go better in the immediate term.  Household consumption or the fight against inflation will be improved.  Impoverishment will be driven underground, progressing in a way almost imperceptible to public opinion. But in ten or twenty years the crisis whose first tremors we are now experiencing will erupt and we will no longer have the resources to recover from it.

The struggle has already started and will be played out in the next few years, but the full implications of the gains and losses will emerge only towards the end of the century.  The drama in Europe is that, however widely it may be shared, this feeling of an unacceptable risk is not of a nature to rally forces.  Men will make sacrifices only if they are actually suffering a crisis or are carried away by enthu-siasm for an ambition or a project.  The artificial content deriving from the

existing balances minimizes the painful effects of the crisis in its initial stages;
the crisis can therefore pursue its hidden course without arousing reactions.  Few
proposals based on hope or feeling have been made.  This study is an attempt to
sketch the broad outlines of a project and to describe ways of making Europe the
laboratory of a new way of life, the centre of a new Renaissance.  In international
competition, Europe would select its own ground on which to fight.  It would no
longer seek only to compete at economic level, but would lead its main rivals
towards the concept of a better life that would be remarkably well suited to its main
specific features.

*    *    *

A Renaissance can take place only on two conditions:  if the forces attaching us
to the past have become negligible or are challenged, and if there is a desire for
the future to be newly constructed to comply with the new ideas that are shaping
our destiny.

The break with the past was not complete as long as the philosophies and ideologies
derived from the previous socio-economic context remained the reference point for
our thinking.  This heritage from the last century, which for long has served as a
freezing and blocking force, is now rapidly vanishing.  It is curious to note that
this reduction in the influence of these ideologies follows very closely on the
heels of the weakening of the system of moral values of "middle class" society.

The manifestations of this deep desire for change are a rather incongruous and con-
tradictory collection:  Vatican II, Club of Rome, May 1968, Solzhenitsyn, Sakharov,
Eurocommunism, the "new philosophers", the ecological movement, the Helsinki agree-
ments, the evolution in the status of women, the Carter Plan for energy, etc.  Some
may view these events as portending the decadence of the white man, chiefly at the
expense of Europe.

This assumption of decadence cannot be ruled out.  However, other signs are more
optimistic as new constructive concepts are emerging.  Ilya Prigogine[1] demonstrated
remarkably well this metamorphosis of the interaction between science and society,
which we are just starting to witness.  His explanation for the behaviour of the
dissipative structures extends to the social phenomena built up on an order that
is tending to change constantly. We are remote indeed from the glorious scientists
whose laws were going to provide explanations for everything and guarantee happiness
in the world.  But the part played by the unexpected, the humility experienced in
the face of the irreversibility of the phenomena of the living world and the unpredictable nature of influential fluctuations are suddenly giving a new meaning to
the concept of freedom.  Freedom is no longer merely an advantage for individuals
or groups that enjoy it, but is above all the condition for adaptation, the necessary slack in the social machinery, to ensure that certain ends can be attained.
Freedom can no longer merely be reduced to a philosophy of human rights; it is now
seen to be the instrument for the uprooting of outdated structures and the establishment of relevant new ones that will also be temporary.

According privilege to the concept of freedom in this way does not mean that man
is powerless, forced to submit to the unpredictable laws of evolution.  As these
concepts were gaining ground, biology and information theory were at the same time
making an essential contribution towards the understanding of complex systems and
their regulation.  As a result, we can now see how to plan without compulsion and

---

[1] 1977 Nobel prize for Chemistry — member of CERD.  "Metamorphose de la Science;
Culture et Science d'aujourd'hui",  paper at the ESIST conference held in Brussels
at the end of May 1977 under the auspices of the Commission of the European
Communities.

how to organize without imposing autocratic hierarchical systems.

An image of these new balances of forces in social organizations is obtained if we observe the hormonal balances in the higher orders of living beings. The cells are immersed in an inland sea consisting of a variety of chemical substances, the interplay of their conflicts and balances governing metabolism. These hormonal messages, loaded with specific information, are remarkably powerful. They act on our growth, our waking or sleeping activity, and our aggressiveness. The exercise of political power must be understood today in the light of these hormonal systems.

At Community level, power will be reflected in the issuing of specific messages of interest to the Community, which will most frequently act by catalytic means, influencing the sensitive points at which systems of value are developed and links of communication and interdependence are established. This is how the confederal ambitions of the Community political system should be understood.

Although it is far too early to declare that a system of this type must be adopted by the Member States in their difficult quest for a balance of forces within the Community authority, it is not too late to try it out in the field of scientific and technical research. And if this attempt were to be made in an effort to foster a Second Renaissance, we should have an opportunity of combining the testing of a new model of political action with the formulation of a project based entirely on the desires of man.

We can give no more than a vague outline of the content of this new Renaissance, as it will be brought about by children who are still at school or not yet born. The minds of these future leaders, unlike our own, will not be shaped by the written tradition and by a universe of mechanical machines, but by the environment of the information civilization into which we have broken almost without realizing it. Their minds will not be the same as ours. During childhood and adolescence, when their sensitivities are being formed, they will be exposed to audiovisual means of dissemination and to the use of programmable computers for games and creative activities in the graphic and plastic arts and in music. However, if we venture to make a prediction, we can base it on what we know of the direction of evolution, the drive towards complexity and psychism. This Second Renaissance will meet man's heartfelt desire if it enriches his spirituality, individually, and collectively. The twenty-first century will tend towards a civilization of the mind, or else it will come up against formidable material and human obstacles.

These few words are sufficient to show the enormity of the effort that must be made to achieve this transition and the great danger of the transitional phase. There are probably many possible solutions and many different approaches to improving the consumption of cultural goods in order to compensate for a restriction in the consumption of material goods. However, all these ways involve progress in the mastery of complexity, in the practical forging of bonds of solidarity ...... and in the acceptance of the constraints of interdependence, in the challenging of the values concerning the use of the time allotted to different pursuits throughout one's life, and, finally, in the more equal distribution of intellectual, artistic, and social culture which is increasingly proving to be the most precious asset of the human race.

**Stimulus**

Revival of economic hope

Aid to countries embarking on the process of industrialization

Increased dissemination of information and knowledge

Promotion of productive spare-time activities

**Ways and means**

Increase in service productivity

New sources of energy and energy-saving

Education with a cultural bias

**Promoting mechanism**

Scientific and technical research acting according to a
procedure of a biological nature controlled by a programme
of priorities

Financing of innovation of a social nature

**Objectives**

Priority for individual or small group activities

Emphasis on cultural pluralism

Printed in Dunstable, United Kingdom